Low Fodm:

CU01461085

Enriched with the Power of the Alkaline Diet To Produce Superior Relief To Digestive Disorders and Acid Reflux To Soothe the Gut for Optimum Health With Shelley Aviv MD Felicia Young Daryl Shrode

James Shepherd
Susan Green Aniys
Redford E. Gordon

Table of Contents

PART 1..3

Chapter 1 Introduction to the Low FODMAP diet.......................4

Chapter 2:Benefits of the Low FODMAP diet............................ 13

Chapter 3: Starting the Low FODMAP diet............................... 19

Chapter 4: Low FODMAP diet foods 36

Chapter 5: Low FODMAP Diet Meal plan 50

Chapter 6: Low FODMAP diet tips and tricks for success................. 78

Chapter 7: Low FODMAP diet FAQ 82

PART 2 85

Chapter 1: What is the Alkaline Diet?................................. 87

Chapter 2: What is a pH Balance? 91

Chapter 3: The Science Behind pH Imbalance 95

Chapter 4: Why Alkaline is Best 99

Chapter 5: Creating an Acid-Alkaline Balance 103

Chapter 6: Alkaline Diet for Vegetarians...........................107

Chapter 7: Alkaline Meal Ideas...................................... 111

PART 3 115

Chapter 1: Chronic Acid Reflux & Its Serious health Implications 117

Chapter 2: The role of Fibre, Prebiotics and Probiotics 124

Chapter 3: Understanding the role of proteins, carbs, AND fats in healing acid damage... 131

Chapter 4: Exercise to reduce acid reflux...........................137

Chapter 5: How to live a reflux free life?...........................139

CONCLUSION...144

PART 1

Chapter 1 Introduction to the Low FODMAP diet

Do you suffer from abdominal cramping and discomfort? If you spend your days feeling constipated, bloated, and feel the uncontrollable urge to use the bathroom? If so, you may suffer from IBS.

With so many diets on the market, it can be hard to decide which one is best for you! In the following chapters, you will be learning everything you need to know about the FODMAP diet and how it can benefit your life.

Unfortunately, there are several theories behind why individuals suffer from IBS. For many, there is 70% of women who suffer from IBS due to their hormones triggering the symptoms. As for others, the reasons could be anything from a sensitive colon, an immune response to stressors, sensitive brain activity in detecting gut contractions, or even a neurotransmitter serotonin being produced in the gut. While the doctors are unable to pinpoint an exact reason for IBS, the good news is that they are certain that IBS will not cause other gastrointestinal diseases and it is not cancer!

The right question to ask in this moment, is what can I do about it? We are here to tell you that the low FODMAP diet is the way to go. In the chapters to follow, you will learn everything from what the diet is, who the diet is for, what FODMAP even stands for, and why this diet will work for you. We cover the benefits of the diet and include an easy start guide so you can get rid of that discomfort and bloat as soon as possible!

To start, it is important to understand what the FODMAP diet is, and why it is something you need to start. However, before we start, here are some tips for the beginners who are just starting or considering the low FODMAP diet.

Getting Started

Before we begin, it is important to get a diagnosis from your family doctor. Many people self-diagnose themselves with IBS and place themselves on the low FODMAP diet. This is something we do not recommend. If you have symptoms such as pain and bloating, you should see a professional to rule out any possible life-threatening diseases.

What is IBS?

As mentioned, be sure to see a professional to attain an official prognosis of IBS. If you suspect you do have Irritable Bowel Syndrome, realize that you are not alone. In fact, around 15% of the population in the United States suffer from IBS symptoms. While the symptoms do vary from person to person, the typical symptoms are as follow:

- Bloating
- Constipation
- Diarrhea
- Lower Abdominal Pain
- Lower Abdominal Discomfort

If you suffer from any of these, it is important to consult with your doctor the specific symptoms you have. This will be vital as there are three different types of irritable bowel syndrome. These include:

- IBS with Constipation
  - Typically, IBS with constipation has symptoms including bloating, abnormally delayed bowel movements, stomach pains, and loose or lumpy stool.
- IBS with Diarrhea
  - Typically comes with symptoms including stomach pain, urgent need to use the bathroom, loose and watery stool
- IBS with alternating Diarrhea and Constipation
  - Due to the fact that there are several types of IBS, this makes it hard to determine a single drug treatment to help with the symptoms. As we mentioned earlier, you need to consult with a professional. Once you have done this and ruled out any other illnesses, it is time to take a look at your diet.

Who is the diet for?

Typically, the low FODMAP diet is meant for individuals suffering from IBS. The diet itself was created as one of the first food-based treatments to help relieve IBS symptoms. The good news is that up to 75% of patients who had IBS experienced symptom relief when they followed the low FODMAP diet. However, the diet is also helpful if you have any of the following:

- Digestive Disorder
- Gastroesophageal Reflux Disease (GERD)
- Crohn's Disease

- Celiac Disease
- Vegan Gut
- Bloating

Once you have determined that the low FODMAP diet could help your symptoms, it is now time to learn what FODMAP even stands for! This is going to be vital information to carry with you through your diet so you understand what you are eating and why your body is reacting the way it does!

We understand that there are many different types of diets out there. Some of you may be wondering, can I follow my current diet and still follow the low FODMAP diet? The answer varies depending on which you follow, and we will try to answer in a simple manner:

- Vegetarian/ Vegan
  - Yes, this diet is more than possible to follow if you are vegan or vegetarian. With a few tweaks, you can find friendly options and still stick to your regular diet!
- Low-Salt
  - If you follow a low-salt diet, this diet is doable for you. However, it will be vital that you learn how to read and follow food labels. Luckily for you, this is information also included in this book!
- Gluten-Free
  - As you will be learning, the FODMAP diet does exclude wheat, which contains gluten. If you are gluten-free, this diet is easy to follow as you most likely will not be able to

have it anyway!

- Kosher
  - If you have to eat kosher, you can still follow this diet. It will be up to you to find certain kosher foods, but after the elimination diet, you will be able to find the foods and still stick to your original diet.

History of the low FODMAP diet

Originally, the low FODMAP diet was developed by a team of scientists at the Monash University located in Australia. The original research was meant to investigate if the diet would be able to control IBS symptoms with food alone. The university established a food analysis program to study FODMAPs in both Australian as well as international foods.

In 2005, the first FODMAP ideas would be published as part of a research paper. In the paper, the hypothesis was that by reducing dietary intake of certain foods that were deemed indigestible, this could help reduce symptoms stimulated in an individual's gut's nervous system.

Over many years, research has shown that certain short-chain carbohydrates such as lactose, sorbitol, and fructose was the cause behind gastrointestinal discomfort. Once the basis of digestion was studied, the low FODMAP diet was created to help with these symptoms.

What does FODMAP stand for?

FODMAPs are typically found in foods that we consume every day. They are in onions, rye, barley, wheat, garlic, milk, fruits, vegetables, and more!

As you can tell from this very small list (don't worry, we will cover more in the chapters to follow), they are in some of our more common foods! This is why it is so easy to feel bloated for some people, without understanding what is causing it! However, before we dive into how this diet works, you will need to understand the acronym FODMAP.

F- Fermentable

O- Oligosaccharies (short chain carbohydrates)

D- Disaccharides (lactose)

M- Monosaccharides (fructose)

A- and

P- Polyols (Sorbitol, xylitol, maltitol, and mannitol)

The reason you may be suffering from IBS or other digestive issues is due to the fact that most FODMAPs have a hard time absorbing into your small intestine. As a result, these FODMAPs are fermented by the bacteria in your small and large intestine in which results in bloating and irregular bowel movements.

While the FODMAPs cause the digestive discomfort, it is important to understand that it is not the cause of the intestinal inflammation itself. In fact, the FODMAPs produce alterations of intestinal flora that help you maintain a healthy colon. This does not change that the symptoms are still uncomfortable.

What may be causing your IBS symptoms could be a fructose malabsorption or a lactose intolerance. As you will be learning in later chapters, as you begin the low FODMAP diet, there will be an elimination phase where you learn what exactly is causing your symptoms and discomfort.

The source of the FODMAP will vary depending on different dietary groups. In more common circumstances they are compromised as the following:

- Oligosaccharies: Fructans and Galacto-oligosaccharies
- Disaccharies- Lactose
- Monosaccharies- Fructose
- Polyols- Xylitol, Mannitol, Sorbitol

Sources of Fructans

In later chapters, we will be going more in-depth on the foods you can and cannot eat. To cover the basics, you should understand where these specific irritants come from. To start, we will go over the source of fructans. These can be found in very popular ingredients including; rye, garlic, onion, wheat, beetroot, Brussel sprouts, and certain prebiotics.

Sources of Galactans

As for galactans, these are primarily found in beans and pulses. It can also be found in certain tofu and tempeh, but this does not mean that vegans and vegetarians cannot follow the low FODMAP diet. It simply means that you will need to find other sources of proteins if you want to follow a plant-based diet. We will be going over this more in the chapters to follow.

Sources of Polyols

Polyols are typically found in stone fruits. These include avocados, apples, blackberries, watermelon, and more. They are also found naturally in certain vegetables and bulk sweeteners.

While this diet may seem to be lacking many of your favorite foods, don't you worry! Due to the wide variety of IBS symptoms, it is unclear which foods trigger certain individuals. This is why the elimination trial will be important before you start the diet. Please remember that everyone is different. While some people see immediate results when they begin the diet, for others, it will take some time.

Effectiveness and Risks of the low FODMAP diet

It is important to understand that the low FODMAP diet is meant for short-term symptom relief. However, long-term diet can have a negative effect on your body. Unfortunately, it can be detrimental to your guy metabolome and microbiota. It is to be taken very seriously that this diet is meant for short periods of time and only under the advice of a professional.

Please understand that if you choose to follow the low FODMAP diet without any medical advice, it is possible the diet could lead to some serious health risks. Some of these risks are as followed:

- Nutritional Deficiencies
- Increased Risk of Cancer
- Death

When you start the low FODMAP diet, it is possible the diet itself could mask any serious disease that present themselves of digestive symptoms.

These could include celiac disease, colon cancer, or inflammatory bowel disease. This is why it is so crucial to seek professional help before starting the diet on your own.

Now that you have learned the basics of the low FODMAP diet, it is time to learn all about the benefits that come with the diet change. Obviously, the main change will be to help lower any digestive troubles you may behaving. By removing the potential triggers in which are causing your digestive issues, this will help pinpoint which food intolerances you have.

While this diet may seem to take a lot of time and effort, think of the time you are wasting by being in discomfort all of the time and using the bathroom! With a few minor adjustments and tests, you will be able to find the source of your problem and hopefully never feel this way again! Now, onto learn all of the other incredible benefits the low FODMAP diet can bring to you!

Chapter 2: Benefits of the Low FODMAP diet

According to research, the low FODMAP diet is effective for around 75% of patients who suffer from IBS. In most cases, the patients are able reduce any major symptoms they are experiencing and in hand, improve their quality of life.

In the same research, scientist found evidence that the diet can also be beneficial for people who suffer from other functional gastrointestinal disorders such as Chron's disease, ulcerative colitis, and inflammatory bowel disease. All you need to do to benefit from this diet is to figure out what is causing the digestive disturbances and symptoms. Below, you will find some of the other benefits the low FODMAP diet has to offer:

A. IBS Symptom Reduction

By following the low FODMAP diet, individuals can reduce most symptoms involved with IBS including stomach pain, bloating, and gas. It is important to follow the diet and remove any irritants as they ferment inside of your intestines. By selecting foods that don't trigger your symptoms, you can avoid them altogether!

B. Chron's Disease Reduced Discomfort

By following the low FODMAP diet, individuals were able to change the quality and number of prebiotics. By controlling the foods you consume and avoiding the ones that trigger your system, you could reduce the discomfort you feel from the trigger foods.

C. Increased Energy

Some individuals feel tired no matter how much they eat through the day. It is believed that a low FODMAP diet can help reduce fatigue. This could be due to the fact that the body is no longer wasting energy on digesting foods that don't agree with your system. This is especially true for sweeteners that you could be using on a daily basis. As you will be learning later, some of the best sweeteners can be found in fruit!

D. Reduced Constipation and Diarrhea

When you follow the low FODMAP diet, you will begin to eliminate foods that are causing your symptoms in the first place. When you do this, your body will find a balance, and you may find that your bloating will decrease, the gas will decrease, and your stools will return to normal. It is a win-win situation! All you will need to do is figure out your triggers (which we cover in the third chapter) and follow the diet!

On top of these incredible benefits, there is also beliefs that the low FODMAP diet can benefit psychological health. Often times, the disturbances of IBS can cause stress to certain individuals, eventually leading to anxiety and depression. When you remove the trigger causing the symptoms by diet, you will be able to improve the quality of your life.

As you will be learning in the chapters to follow, the low FODMAP diet includes an elimination diet in order to get started. As you introduce foods, you may find that you have a lactose or gluten intolerance. While this may seem like a huge change, there are some incredible benefits to changing your diet.

Benefits of a Grain Free Diet

As you will be learning later in the book, there are many types of grains that has been found to cause inflammation. Unfortunately, this is a very common culprit for the digestive disorders you may be experiencing. This is especially true if you are very sensitive to gluten. The good news is that if you are looking to lose weight on top of feeling better, cutting out these grains will be the best thing to ever happen to you. Reined grains are high in carbohydrates and calories; they offer little to no nutrition and contribute to discomfort in your stomach. Before you make the switch to going grain free, consider some of the amazing benefits as follow:

A. Digestion Benefits

Gluten is a type of protein that can be found in wheat products. If you do the elimination diet and find that you are gluten sensitive, cutting it out makes the most sense. When you cut it out of your diet, this can help relieve issues such as nausea, bloating, diarrhea and constipation.

B. Reduced Inflammation

When you experience acute inflammation, this normally means that your immune system is fighting off foreign invaders. Unfortunately, if you sustain these levels for a long period of time, this is what causes chronic disease. By cutting gluten out, you can reduce the amount of inflammation in your body.

C. Balanced Microbiome

By following a grain-free diet, you will be able to balance the microbiome in your gut. When you do this, it helps support the beneficial bacteria in your body, helping improve your digestion, boost your immunity, and helps keep blood sugar under control.

D. Weight Loss

As mentioned earlier, most grains offer little to no nutrition. When you cut these extra calories out of your diet, it will help you lose weight. Instead of grains, try eating nutrient-dense foods like vegetables or legumes. Of course, you will figure out exactly what you can eat on the low FODMAP diet after going through the elimination portion.

Another common irritant when individuals suffer from IBS and other digestive orders can be lactose! You may be thinking to yourself; I could never give up my yogurt or ice cream. The good news is that in the current market, some incredible alternative choices can fit in the low FODMAP diet. In case you need some further benefits to help convince you, here are just a few:

E. Healthier Digestion

You may not know, but around 70% of the population has a degrees of intolerance to lactose. When we first begin to wean off of our mother's milk, we begin to use lactase. Lactase is an enzyme that helps digest lactose found in milk. As we age, we begin to lose the ability to digest lactose and is one of the biggest known triggers for IBS. By taking dairy out of your diet, you save yourself the troubles all together!

F.   Decreased Bloating

Bloating occurs when we have issues with digestion. Some dairy products can cause excessive gas in the intestines, which is what causes the bloating in the first place. Some bodies are unable to break down the carbohydrates and sugar fully which in turn, creates an imbalance of gut bacteria.

G. Clearer Skin

If you suffer from acne, dairy could be the culprit! According to studies posted in Clinics in Dermatology, it was found that dairy products such as milk contain growth hormones that stimulate acne. By following the low FODMAP diet and cutting dairy from your diet, you could naturally treat acne!

H. Reduce Risk of Cancer

A 2001 study at Harvard School of Public Health found that there was a connection between high calcium intake and increased risk of prostate cancer. It is thought that the hormones in the milk contain contaminants such as pesticides that have been linked to cancer cell growth. These contaminants are mostly found in dairy products, giving you another reason to cut them from your diet altogether!

I.   Decreased Oxidative Stress

It is believed that a high milk intake is typically associated with higher mortality rates in both men and women. This may be due to the D-galatose found in milk which helps influence oxidative stress and inflammation in the body. Unfortunately, this undesirable effect caused by milk can cause chronic expose and damage health. On top of inflammation, it can also shorten life spans, cause neurodegeneration and also decrease one's immune system.

As you can tell, there are so many incredible benefits of switching over to the low FODMAP diet. Whether you are looking to get rid of bloating, lose some weight, or stop constipation and/or diarrhea, the low FODMAP diet has got you covered. It is all a matter of figuring out what your trigger is in the first place.

Obviously, we could go on and on about the incredible benefits of the diet, but then we would never get to the diet itself! Now that you are aware of just some of the benefits, it is time to get you started! In the chapter to follow, you will be learning how to get started on the low FODMAP diet. You will the steps to get started on the diet itself and how to diet whether you are vegan, vegetarian, diabetics, or doing this for a child who suffers from IBS. When you are ready, we can dive in!

Chapter 3: Starting the Low FODMAP diet

Now that you are aware of the low FODMAP diet and some of its benefits, it is time to learn how you can get started on the diet yourself! While a diet and lifestyle change can seem daunting, it will be important to believe in yourself and remember why you started it in the first place. In the chapter to follow, we will be providing you with all of the information you need. From diagnosis of IBS, to starting the diet, and even how to practice if you are vegan, vegetarian, or diabetic. This diet can be universal; it is all about finding what works best for you. First, it is time to understand the diagnosis of IBS.

Getting Diagnosed with IBS and FODMAP Tests

If you are in the process of being diagnosed with a chronic medical condition, this could be a challenging time for you. It is important that you understand the symptoms the doctors are looking for, and which medical tests you will be taking in order to be officially diagnosed with irritable bowel syndrome. IBS can be diagnosed with a combination of Rome IV criteria so the doctors will be able to rule out any other gastrointestinal disorders.

Rome IV is a set of criteria that doctors have found that most IBS patients have in common. This criteria is 98% accurate when the doctors are identifying their patients with IBS. These criteria are as followed:

1. Recurrent abdominal pain at least one day per week in the last three months are have the following:

   A. Related to defecation

B. A change in stool frequency

C. Change in form of stool

2. Criteria from above is fulfilled with symptoms for at least six months before official diagnosis

Other symptoms often associated with IBS include bloating, abdominal pain, and a change in bowel habit. Your doctor will take in the evidence and match with the Rome IV criteria and will move onto discussing any red flag symptoms that may be occurring.

Before being diagnosed with IBS, it will be important that your doctor rules out any other medical conditions that could be presented with the same symptoms; this is where the red flag symptoms come into play. These flags include:

- Inflammatory Markers
- Rectal Masses
- Abdominal Masses
- Anemia
- Nocturnal Symptoms such as waking up from sleep to defecate
- Family History of coeliac disease, inflammatory bowel disease, and ovarian cancer
- Rectal Bleeding
- Unintentional Weight Loss

On top of these symptoms, your doctor will also ask for several other symptoms in order to diagnose you with IBS. Firstly, the discomfort and pain in your abdomen will need to be related to altered bowel frequency as well as a change in your stool form. You will also need to have at least two of the following symptoms:

1. Feeling incomplete emptying when using the bathroom
2. Passage of mucus when using the bathroom
3. Straining, Urgency, and Altered Stool passage
4. Abdominal Bloating
5. Lethargy
6. Backaches
7. Bladder Symptoms
8. Nausea

Medical Tests for IBS

Once your doctor figures out your symptoms, rules out any other serious medical conditions and believes it is appropriate to run tests for IBS can you expect one of the following tests:

- Antibody testing for coeliac disease
- C-reactive protein
- Erythrocyte sedimentation rate
- Full blood count

While these tests are typical, it may be different if you present any of the red flag symptoms from above. If you do have a red flag symptom, there will be additional tests to rule out any more serious issues. These tests are as follow:

- Hydrogen Breath Test (meant for lactose intolerance)

- Fecal Occult Blood

- Fecal Ova and Parasite Test

- Thyroid Function Test

- Rigid/ Flexible Sigmoidoscopy

- Ultrasound

In the case that your doctor feels your symptoms may not be linked to IBS, they will most likely refer you to a gastroenterologist. This is a physician who is an expert in managing diseases found in the liver and gastrointestinal tract. However, this is worst case scenario. For now, we will focus on following the diet if you are diagnosed with IBS.

Breaking down FODMAPS

In general, FODMAPs naturally occur in popular foods such as vegetables, fruits, grains, cereals, dairy products, and legumes. Unfortunately for those who suffer from IBS, these FODMAPs are absorbed poorly in our small intestines and can affect our bowels as a symptom. FODMAPs are short-chain carbohydrates found in these foods, but this does not mean that the diet itself is sugar-free. When we consume FODMAPs, they are fermented by gut bacteria in the large intestine in which triggers the unpleasant GI symptoms that you may be experiencing. Before we move onto the elimination stage, it is important to understand just what this acronym means.

Fermentable

Fermenting is the process where our gut bacteria attempt so break down

FODMAPs. As you are already aware, these FODMAPs are indigestible carbohydrates and in turn, produce gas.

Oligo-Saccharies

This group of the FODMAP are broken down into two subgroups including fructans and galactans. Fructans also known as fructo-oligosaccharies or FOS are most commonly found in foods such as dried fruit, barley rye, wheat, garlic, onion. Galactans or galacto-oligosaccharies or GOS are found in pulses, legumes, cashews, pistachios, and silken tofu. If you feel yourself panicking over remembering these foods, don't worry. In the chapter to follow, we will cover exactly what you can and cannot eat while on the low FODMAP diet.

Di-Saccharies

As mentioned in the chapter from before, lactose could be a potential trigger in your diet. These can be found in any product that comes from goat, sheep, or cow's milk. Lactose itself contains two sugars united that require an enzyme known as lactase before our bodies are even able to absorb it. When your gut lacks these enzymes, this is when you can trigger symptoms of IBS.

Mono-Saccharides

This is a fructose that is found when a person has an excess amount of glucose in their diet. Our bodies need an equal amount of glucose in our system to stop any malabsorption. This means that while some of us can consume a certain amount of glucose, it is important to avoid foods that contain an excess amount. Some examples of these excessive foods include asparagus, honey, apples, and pears.

And Polyols

Polyols are also known as sugar alcohols. These can be found in a wide range of vegetables and fruits including sweet potatoes, mushrooms, pears, and apples. These sugar alcohols are also found artificial sweeteners in chewing gum, diabetic candy, and even protein powder. These polyols can only be partially absorbed into our small intestines. The rest continue into the large intestine, begin to ferment, and cause discomfort and bloating for some people.

As you begin to consider the low FODMAP diet, it is important to understand that one size does not fit all. This diet will change depending on your intolerance to certain foods. On the FODMAP diet, you will be following three different phases including: The Elimination Phase, The Reintroduction Phase, and the Maintenance Phase. We will go further into detail of each phase, so you have a full understanding before beginning.

The Elimination Phase

The Elimination phase is also known as the restriction phase. While this may seem intimidating, realize that this phase is only meant to last two to six weeks. This phase should only last long enough for you to gain control over your symptoms. Once this happens, you will move onto the reintroduction phase with the help of a professional. It is important that this stage is short as it can have long-term effects on your gut health.

To begin, you will want to create a personal list of foods you feel makes your IBS worse. If you are unsure which foods could be causing your symptoms, you will want to check out the next chapter to see an extended list of foods you should be avoiding. Some popular starters include chocolate, coffee, nuts, and certain fibers.

Once you have made your list, you will begin to eliminate these foods one at a time from your diet. It will take a couple of weeks before you notice any improvements. It does take some time for these foods to get through your system. However, if you do not notice any improvement, you will want to reintroduce these foods into your diet and try the next item on your list. Eventually, you will have a complete list of foods that trigger your IBS symptoms. Other popular foods to eliminate during this phase include: soy, gluten, and dairy products.

An important tool during this phase will be a food diary. This way, you will be able to keep track of which foods you are eating during the day, and any symptoms that may present themselves after they have been consumed. In general, the longer this phase is, the more likely you are to find that is triggering your IBS symptoms. It is important to remember that once eliminated, you will need to reintroduce foods slowly in the next phase.

The Reintroduction Phase

Once you have gone through your elimination period, you will be reintroducing these targeted foods back into your diet. While following the low FODMAP diet, you will need to introduce these foods back into your lifestyle one at a time.

As a tip for the reintroduction phase, we suggest starting on a Monday. This way, you will be able to consume a small portion of the food, wait a few days, and see if you experience any symptoms. On day three, you can eat a larger portion and wait another couple days for any onset symptoms. Be sure to keep track of how you are feeling in your food diary so you can present it to a professional if need be. If you experience symptoms, this is a possible food trigger. If there is no symptom, you can assume that this certain food group is a good match for your diet.

After a while, you will have a complete list of foods that you need to assess, and you will start the elimination phase over again to double check. Once you eliminate and reintroduce, you will be able to create a diet you can stick with and eliminate any symptoms of IBS you may experience.

Maintenance Phase

While this may take time, the elimination and reintroduction phase are going to be very important while following the low FODMAP diet. These are going to be your tools in identifying foods that trigger your IBS symptom. The long-term goal is to create a wide variety of foods you can consume on a daily basis to ensure you are intaking all of your essential nutrients while eliminating the ones that make you feel lousy.

As you go through these phases, it is vital you listen to your body. Only you will be able to tell if you have a tolerance to certain foods. Remember that portion sizes will be important during these phases as well. While you may not react to small portions, larger portions may trigger the symptoms which you will want to avoid. The more tests you do, the more foods you will be able to add or subtract from your diet. While this may take some extra work, it will be worth it when you decrease your bloating and discomfort from IBS.

You may be wondering if you can follow this diet even if you have a certain lifestyle. Typically, the answer is yes! The only factor being that you could have a very limited number of foods allowed on your diet with any other limitations. Below, we will cover some of the more common diets and how you can also follow the low FODMAP diet as well!

Low FODMAP Diet with Vegan/Vegetarian Diet

If you follow a vegan or vegetarian diet, you may want to consider working with a dietary professional. Due to the fact you consume a diet that is different from most of the population, it can be more difficult to access foods that can work well with both diets. By working with a professional, they can ensure you still follow your diets without missing any essential nutrients your body needs.

While on the low FODMAP diet, it is important you keep re-testing foods. Remember that the elimination phase is meant to be short term. As you reintroduce old foods, you will be able to process if you can tolerate them or not. While you do this, you can find some staple foods, even if they happen to be high in FODMAPs.

If you follow a vegan or vegetarian diet, it will be vital that you pay special attention to your protein intake. As you will be learning later in the book, the low FODMAP diet includes a limitation of many legumes which may be a main source of protein for you right now. Instead of legumes, you can consider soy products or simply a smaller portion of legumes. Along with these switches, there are also milk substitutes to help with your protein intake. There is almond milk and other soy products to help out. Certain nuts and seeds also have varying levels of proteins for you to consider.

Low FODMAP Diet and Diabetes

If you have diabetes, you are most likely aware that there is no specific diabetic diet. In general, most people with diabetes follow a suggested balanced and healthy diet. If you wish to follow the low FODMAP diet while having diabetes, there are a few key rules you can follow to ensure you do not cause further harm to your health.

1. Planning

While on the low FODMAP diet, planning regular meals will be key. By doing this, you will be able to make sure that your blood glucose levels are always stable. By planning in advance, you can be successful in managing your diabetes while still following the low FODMAP diet. This stands especially true if you struggle finding healthy foods when you are away from your house. By being prepared, you will always have healthy options and can stay away from temptations. One good idea is to prep snacks for in-between your meals. These can be rice cakes, popcorn, or a simple fruit that is allowed in your low FODMAP diet.

2. Focus on Low FODMAP Carbohydrates

If you wish to follow the low FODMAP diet and eat healthy while having diabetes, eating starchy carbohydrates will be important for you. Some suitable options for you include wheat free bread, oats, potatoes, and rice. Before you include these, be sure to eliminate them from your diet to assure they are not triggers. Essentially, you will want to avoid any large portions of carbohydrates so you will be able to avoid any spikes in your blood glucose. You can do this by choosing slow-release carbohydrates like sweet potatoes or oats. On top of these carbs, you will also want to include allowed vegetables.

3. High Sugar Foods

As a person with diabetes, you already know that sugary foods cause your blood glucose levels to rise. On the low FODMAP diet, there is a low risk of consuming sugary foods such as soft drinks and cake, but they should still be avoided.

4. Low FODMAP Fruit

While fruits are a source of sugar, it will be important that you include a few portions of fruit per day. On the low FODMAP diet, there are plenty of options such as grapes, strawberries, bananas, and even oranges. You will want to pay special attention to your portion sizes as bigger portions, means higher amounts of fructose. You will also want to limit your portions of dried fruit, smoothies and fruit juices as they are typically pretty concentrated sources of fructose.

Low FODMAP Diet for Children

At this point, there has been very little research on the low FODMAP diet for children. Studies have shown that there are no real negative side effects for individuals who follow the low FODMAP diet for short period of time. However, if this diet were to carry on for longer than suggested, it could possibly have a negative effect on the gut flora balance in a child. If you are considering this diet for your child, there are several factors you will want to take into consideration.

First, your child will need to be seen by the pediatrician to confirm that your child has irritable bowel syndrome. Once it is diagnosed, the doctor will need to approve the diet and be carefully supervised to assure the safety of your child. Only after you follow these steps should you continue onto the elimination stage of the low FODMAP diet. For success on the diet, you can follow some of the following tips:

1.   Inform Other Adults

Just like with any other diet restrictions, you will want to inform key adults of your child's restrictions. Whether it is a friend, a child care provider, or a teacher, this will be vital for the success of the diet. When these adults are in know of your child's diet, they will be able to address any stomach issues they may be having.

2.   Involve Your Child

If your child is old enough, try to explain the diet to them in simple terms. You will want to explain that they are feeling sick due to the food they are eating. Be sure to include them and ask for their input in the food substitutions and menu. By making your child feel they are a part of the process, this may help your child comply with the new food rules.

3.   Pack and Plan

Many parents fear diets for their children as they are always on the go! Luckily, the FODMAP diet is pretty easy to follow when you plan ahead. When you are at home, you most likely stock the fridge with low FODMAP foods. By planning ahead and packing your own snacks and lunches, you can assure your child will stick to the diet, so they do not make themselves sick.

4.   Forget the Small Stuff

Your kid is going to be a kid. If your child eats a restricted food every once in a while, it isn't going to ruin their diet altogether. Children typically do not have the self-discipline that adults have. They will most likely be tempted by restricted foods when at school or with their friends. You need to remember that while you want to stick to the diet most of the time, you can still allow your child some freedom when it comes down to what they are eating.

Exercise on the Low FODMAP Diet

While your diet may be causing your IBS symptoms, research has found that exercise can also help decrease any symptoms you may be suffering from. There are a few reasons why including regular moderate exercise will be important in the success of your diet.

First off, regular exercise can help reduce stress in your body. Typically, IBS tends to stress people out. When this happens, the nerves in your colon become tenser and can create abdominal pain. When your colon is tenser, this can slow down your bowel movements all together and cause constipation. A simple exercise such as cycling or walking can help release endorphins into your system and help release the tension in your colon. The more relaxed you are, the more flexible you will become.

Along with decreased stressed will come an increase of oxygen in your body. There are plenty of wonderful exercises such as tai chi and yoga that creates a breathing routine. When you take in these abdominal breaths, this helps increase the amount of oxygen in your body. As you increase oxygen, this will also help release any tension you are holding in your colon.

Finally, exercise can also increase your blood flow. As you begin to sweat, your body will be getting rid of toxins that could be creating discomfort in your colon. The more you sweat, the healthier you will be. Plus, the movement could help promote healthier bowel movements by moving blood to any problematic areas you may have.

As you consider exercise with your diet, remember that it will be vital to fuel your body before and after exercise. You will want to fuel about one to two hours before you work out. As long as it is included in your low FODMAP diet consider a banana with peanut butter or even oatmeal with some strawberries. The exercise can be any moderate activity of your choice from dancing, to running, to cycling, or even a little bit of strength training. Choose an exercise that makes you happy and one that you will stick with.

Reasons the Diet May Not Be Working

Speaking of sticking to a diet, some of you may follow these instructions and still suffer from IBS symptoms. If this still happens, you will want to take a look at your stress levels and the diet itself. While of course there is going to be a learning curve of the low FODMAP diet, allow yourself several weeks to change your food habits. Feel free to check back to the resources of this book to assure you are eating the foods allowed on the low FODMAP diet. If you still have no idea why you are experiencing the symptoms still, perhaps it is one of the following reasons that the diet isn't working:

1.   Lack of Fiber

Fiber plays a very important role in keeping your stool regular. Often times, the low FODMAP diet will remove high fiber foods, which means you will need to pay special attention to your fiber intake. If you find yourself struggling, try speaking to a professional to find other options while on the low FODMAP diet. It will also be important that you drink plenty of water to move fiber through your system.

2.   Too Much Fruit

While there are plenty of fruits on the low FODMAP diet, it is possible you are eating too much of it in one sitting. Typically, you will want to stick to only one serving at a time. If you want more fruit later in the day, try waiting two to three hours after the first one is consumed. As **you** practice this diet more, you will be able to tell your tolerance levels with the fruits so you can reduce that time in between servings.

3. Hidden FODMAPs

Often times, you could be consuming ingredients that are high in FODMAPs and have no idea. Typically, they are found in highly processed foods to help their taste and texture. FODMAPs are also found in some medications such as cough drops and cough syrup. Unfortunately, while they can help a cold, they are often high in sugar alcohols which can trigger your IBS symptoms. It will be important to read labels, which is included in the chapter to follow.

4. Portion Control

It is very easy to sit down and eat more than a portion. While on the low FODMAP diet, allowed foods can become high FODMAPs when you exceed their allowed portion size. As an example, you may want to enjoy some rice cakes as a treat. A recommended serving size is only two rice cakes. If you eat double the allowed portion, this is when you may experience symptoms of IBS. Again, this is where reading labels carefully will come in handy while on the low FODMAP diet.

5. Stress

Stress is going to be a huge factor on the low FODMAP diet. If you are carefully following your diet, check your lifestyle. Stress itself can cause IBS symptoms so you may want to consider stress management skills along with a diet. You can try therapy or yoga. At the end of the day, your success is in your own hands.

If you continue to have IBS symptoms after following the diet and dealing with the issues from above, you may want to seek medical advice again. It is possible you have further intolerances that have not been explored yet. Also, the FODMAP diet will not work for everyone. If you have tried and failed, ask your doctor or dietician what the next step for you could be. For now, we will begin to cover the foods you can and cannot eat while on the low FODMAP diet.

Chapter 4: Low FODMAP diet foods

In the chapter to follow, you will find a list of both low and high FODMAP foods. As for the elimination phase, you will want to try to eliminate all of the high FODMAP foods. Once you are in the reintroduction stage, you will be able to introduce these foods back in order to see what is triggering your IBS symptoms.

As you choose your foods for your low FODMAP diet, remember that reading the ingredient list on a package is going to be vital for your diet success. Below, we will cover some of the basics of reading a food label. Too often, companies are able to hide food ingredients and could be triggering your symptoms without understanding why.

When you choose your foods, portion control will also be vital. When it comes to fruit, try your best to portion out one piece every few hours. As for processed foods, you will want to avoid them all together. If you ever have any doubts on low and high FODMAP foods, you can always revisit this chapter!

Reading and Understanding Nutrition Fact Label

If you are looking to eliminate certain foods from your diet, you will be surprised to learn that they can sneak into dishes without even realizing they are there. In order to stick with your diet, learning how to read and understand a nutrition fact label is going to be crucial for your diet.

A. Serving Size

When you first look at a label, you will want to check out the serving size along with the number of servings in any given package. These serving sizes are typically standardized so you can compare them to other similar foods. Remember that for some people, they can have smaller portions of FODMAP foods, but bigger portions could trigger IBS symptoms. When you are aware of a true serving size, this will make sticking to your diet a bit easier.

B. Calories

If you are on the low FODMAP diet to lose weight, this could be helpful for you. The calories in each package provide a measurement of how much energy comes in a serving of the food. The more calories you consume, the more you will gain weight. By being mindful of the calories in a portion, you will be able to manage your weight in a healthy manner.

C. Nutrients

When you look at a label, the first ones listed are typically the ones that Americans eat a good amount of. These can include Total Fat, Saturated Fat, Trans Fat, Cholesterol, and Sodium. While this isn't the main focus of the low FODMAP diet, it is something you should be mindful of for your general health.

D. Ingredients List

Finally, you will want to pay special attention to the ingredients list included on the package. If you are intolerant to certain ingredients, you will want to keep a food journal of these foods, so you always have them at hand to compare to a label. When looking at the ingredients list, they will be listed in order of weight from most to least. Eventually, you will know exactly what you can't eat and be able to compare easily in the store. As a beginner, remember to read the label of everything you put in your shopping cart.

When you understand the basics of reading a label, it is time to move onto learning the high and low FODMAP food list. We will begin with the high FODMAP foods. With this list, you will either want to avoid the foods altogether, or reduce them drastically. Of course, everyone's tolerances will be different but to help reduce any symptoms of IBS, you should reduce the following foods to enhance your health.

High FODMAP Foods (Avoid/ Reduce)

Fruits (High Fructose)

- Apples
- Avocado
- Apricots
- Blackcurrants
- Blackberries
- Boysenberry
- Currants
- Cherries
- Dates
- Figs
- Feijoa
- Guava
- Grapefruit
- Goji Berries
- Lychee
- Mango
- Nectarines
- Prunes
- Pomegranate
- Plums
- Pineapple
- Persimmon
- Pears
- Peaches
- Raisins
- Sultana
- Tamarillo
- Watermelon

Vegetables/ Legumes

- Asparagus
- Artichoke
- Butter Beans
- Broad Beans
- Black Eyed Peas
- Beetroot
- Bananas
- Baked Beans
- Choko
- Celery

- Cauliflower
- Cassava
- Fermented Cabbage
- Garlic
- Kidney Beans
- Leek
- Lima Beans
- Mushrooms
- Mixed Vegetables
- Pickled Vegetables
- Peas
- Red Kidney Beans
- Soy Beans
- Shallots
- Scallions
- Split Peas

Cereals and Grains

- Almond Meal
- Amaranth Flour
- Breadcrumbs
- Bread
- Biscuits
- Barley
- Bran Cereals
- Crumpets
- Croissants
- Cakes
- Cashews
- Cereal Bars
- Couscous
- Egg Noodles
- Freekeh
- Gnocchi
- Muesli Cereal
- Muffins
- Pastries
- Pasta
- Pistachios
- Udon Noodles
- Wheat Bran
- Wheat Cereals
- Wheat Flour
- Wheat Germ
- Wheat Noodles
- Wheat Rolls
- Spelt Flour

Sweeteners/ Condiments

- Agave
- Fruit Bar
- Fructose
- Hummus
- Honey
- High Fructose Corn Syrup
- Jam
- Molasses
- Pesto Sauce
- Relish
- Sugar-Free Sweeteners (Inulin, Isomalt, Lactitol, Maltitol, Mannitol, Sorbitol, Xylitol)
- Tahini Paste

Drinks

- Beer
- Coconut Water
- Fruit Juices (Apple, Pear, Mango)
- Kombucha
- Malted Drink
- Quinoa Milk
- Rum
- Soy Milk
- Soda
- Tea (Black Tea, Chai Tea, Dandelion Tea, Fennel Tea, Chamomile Tea, Herbal Tea, Oolong Tea)
- Whey Protein
- Wine

Dairy

- Cheese (Cream, Halloumi, Ricotta)
- Custard
- Cream
- Ice Cream/ Gelato

- Kefir
- Milk (Cow, Goat, Evaporated Milk, Sheep)

- Sour Cream
- Yogurt

While this may seem like a large list of foods you shouldn't eat, remember that ingredients will affect individuals a little differently. While you should limit the foods listed from above, it is okay to have them every once in a while. The point of this diet is to help reduce symptoms from IBS and bloating. At the end of the day, you are in charge of what you eat and understand how certain foods will make you feel.

Low FODMAP Foods

Fruits

- Ackee
- Breadfruit
- Blueberries
- Bilberries
- Bananas (Unripe)
- Clementine
- Cranberry
- Cantaloupe
- Carambola
- Dragon Fruit
- Guava (Ripe)
- Grapes
- Honeydew
- Kiwi Fruit
- Lime
- Lemon
- Mandarin
- Orange
- Plantain
- Papaya
- Passion Fruit
- Rhubarb
- Raspberry
- Strawberry
- Tangelo
- Tamarind

Vegetables

- Alfalfa
- Butternut Squash
- Brussel Sprouts
- Broccolini
- Broccoli
- Bok Choy
- Beetroot
- Bean Sprouts
- Bamboo Shoots
- Cucumber
- Courgette
- Corn

- Choy Sum
- Cho Cho
- Chives
- Chili
- Chick Peas
- Celery
- Carrots
- Cabbage
- Eggplant
- Fennel
- Ginger
- Green Pepper
- Green Beans
- Kale
- Leek Leaves
- Lentils
- Lettuce
- Olives
- Okra

- Pumpkin
- Peas (Snow)
- Parsnip
- Red Peppers
- Radish
- Sweet Potato
- Swiss Chard
- Sun-Dried Tomatoes
- Squash
- Spinach
- Spaghetti Squash
- Seaweed
- Scallions
- Turnip
- Tomato
- Water Chestnuts
- Yams
- Zucchin

Meat and Poultry

- Beef
- Chicken
- Deli Meats

- Lamb
- Prosciutto
- Pork

- Turkey
- Processed Meats

Seafood and Fish

- **Fresh Fish (Cod, Haddock, Salmon, Trout, Tuna, Canned Tuna)**
- **Seafood (Crab, Lobster, Mussels, Oysters, Shrimp)**

Breads, Cereals, Grains, and Nuts

- Bread
  - Wheat Free
  - Gluten Free
  - Potato Flour
  - Spelt Sourdough
  - Rice
  - Oat
  - Corn
- Pasta
  - Wheat Free
  - Gluten Free
- Almonds
- Biscuit (Shortbread)
- Buckwheat (Noodles, Flour)
- Brazil Nuts
- Brown Rice
- Crackers
- Corn Tortillas
- Coconut Milk
- Cornflakes
- Corncakes
- Crispbread
- Corn Flour
- Chips (Plain)
- Mixed Nuts
- Millet
- Macadamia Nuts
- Oatcakes
- Oats
- Oatmeal

- Pretzels
- Potato Flour
- Popcorn
- Polenta
- Pine Nuts
- Pecans
- Rice
    White
    Rice
    Brown
    Basmati
- Rice Krispies

- Rice Flour
- Rice Crackers
- Rice Cakes
- Rice Bran
- Seeds
    Sunflower
    Sesame
    Pumpkin
    Poppy
    Chai
- Tortilla Chips
- Walnuts

Condiments, Sweets, and Sweeteners

- Almond Butter
- Acesulfame K
- Aspartame
- Chocolate
    White
    Milk
    Dark
- Erythritol
- Fish Sauce
- Glycerol
- Glucose

- Golden Syrup
- Jelly
- Ketchup
- Mustard
- Miso Paste
- Mayonnaise
- Marmite
- Marmalade
- Maple Syrup
- Oyster Sauce

- Peanut Butter
- Rice Malt Syrup
- Sucralose (Sugar)
- Stevia
- Sweet and Sour Sauce
- Shrimp Paste
- Saccharine
- Tomato Sauce
- Tamarind
- Vinegar
  - Rice Wine Vinegar
  - Balsamic Vinegar
  - Apple Cider Vinegar
- Worcestershire Sauce
- Wasabi

Drinks

- Alcohol (Wine, Whiskey, Gin, Vodka, Beer)
- Coffee
- Chocolate Powder
- Protein Powder (Whey, Rice, Pea, Egg)
- Soya Milk
- Sugar-Free Soft Drinks
- Water

Dairy/ Eggs

- Butter
- Cheese (Swiss, Ricotta, Parmesan, Mozzarella, Goat, Fetta, Cottage, Cheddar, Camembert, Brie)
- Eggs
- Milk (Rice, Oat, Macadamia, Lactose-free, Hemp, Almond)
- Swiss Cheese
- Soy Protein
- Sorbet
- Tofu
- Tempeh

- Yogurt (Goat, Lactose-free,   Greek, Coconut)

Herbs and Spices

- Bay Leaves
- Basil
- Curry Leaves
- Coriander
- Cilantro
- Fenugreek
- Lemongrass
- Mint
- Oregano
- Parsley
- Rosemary
- Sage
- Thyme
- Tarragon
- All Spice
- Black Pepper
- Chili Powder
- Cardamom
- Curry Powder
- Cumin
- Cloves
- Five Spice
- Fennel Seeds
- Nutmeg
- Saffron
- Turmeric
- Avocado Oil
- Coconut Oil
- Canola Oil
- Olive Oil
- Sesame Oil
- Sunflower Oil
- Soy Bean Oil
- Vegetable Oil
- Baking Soda
- Baking Powder
- Cocoa Powder
- Ghee
- Gelatin
- Lard
- Salt
- Yeast

As you can tell from the list from above, there are food choices for all different types of diets. Whether you are vegan, vegetarian, or follow a typical diet, there are plenty of choices for you.

The list from above may seem daunting, but as you learn your own version of the low FODMAP diet, you will be able to put together recipes from the ingredients you are allowed. The key to being successful on this diet is enjoying the foods you are allowed. Luckily in today's market, there are plenty of substitutes for ingredients that may trigger you. As long as you take the time to make this list, you will be able to make your new diet successful.

In the chapter to follow, we will be providing a couple different meal plans for you to follow. There will be a seven-day example vegan diet. Once you have read through this, you can move onto the fourteen-day low FODMAP starter diet. Remember that these are mere suggestions and you can make adjustments as needed.

Chapter 5: Low FODMAP Diet Meal plan

At this point in the book, you hopefully have a better understanding of the foods you can and cannot eat while on the low FODMAP diet. Before we jump into potential meal plans for you to follow, it is time to learn some delicious ingredients.

If you feel nervous about the diet due to the big list of foods to avoid, you absolutely shouldn't! Is your diet going to be different? Yes. However, when you are no longer experiencing diarrhea, constipation, bloating, and the other symptoms from IBS, you will be asking yourself why you didn't start sooner!

As you will find out from the recipes from below, there is a way to stick to your diet and enjoy your meal at the same time. You will find easy to make breakfast, lunch, and dinner recipes. Remember to pay special attention to the ingredients so you can determine if the recipe itself will stick within your own limits.

Low FODMAP Breakfast Recipes:

Small Banana Pancakes

Prep Time: Five Minutes

Cook Time: Twenty Minutes

Servings: Two

Portion: Four Mini Pancakes

Ingredients:

- Dairy-free Spread (Olive Oil) (3 T.)
- Ground Nutmeg (.25 t.)
- Ground Cinnamon (.50 t.)
- Salt (.125 t.)
- Baking Powder (.25 t.)
- Brown Sugar (1 T.)
- Gluten-free All-Purpose Flour (2 T.)
- Egg (2)
- Banana (2 Small, Unripe)

Instructions

1. Begin by heating a medium pan over medium heat before tossing in your dairy-free spread.
2. While this is cooking, go ahead and peel the banana before placing it into a bowl. Mash the banana until it becomes smooth and then add in the egg.
3. Once the egg and banana are mixed well, go ahead and add in the rest of the ingredients. At this point, you should have a mixture that resembles batter.
4. Spoon the mixture into your heated pan and cook the pancakes for a few minutes on each side or until they turn a nice golden color.
5. For extra flavor, try topping the pancakes with your favorite low FODMAP fruit!

Roasted Sausage and Vegetable Breakfast Casserole

Prep Time: Twenty-Five Minutes

Cook Time: Forty-Five Minutes

Servings: Eight

Ingredients:

- Eggs (12)
- Low FODMAP Milk (.50 C.)
- Dried Oregano (.50 t.)
- Salt and Pepper (.25 t.)
- Leek Tips (.50 C.)
- Red Bell Pepper (1)
- Lamb Sausage (1 Package)
- Baby Spinach (2 C.)
- Potato (1)
- Butternut Squash (1)
- Sweet Potato (1)
- Olive Oil (1 T.)

Instructions:

1. Before you begin prepping your food, you will want to preheat your oven to 400 degrees.
2. As your oven heats up, prepare the vegetables from the list above by peeling them and dicing the ingredients into bite-size pieces.

3. Once this is done, place the vegetables on a tray and drizzle them lightly with olive oil or a spread that is allowed on your own low FODMAP diet. Pop them into the heated oven for twenty minutes or until they are soft.

4. While the vegetables are cooking, you can cook your red bell pepper, leek, and sausage in a pan over medium heat. Be sure to cook all of these ingredients through.

5. Now that all of these ingredients are cooked, add in the vegetables to a large casserole dish.

6. In a small bowl, mix together the eggs and add in desired spices. When ready, gently pour the mix over the vegetables already placed in the casserole dish.

7. Place the dish in the oven for thirty minutes or until the eggs are set. This is a great dish to enjoy hot or cold for breakfast!

Blueberry Low FODMAP Smoothie

Prep Time: Five Minutes

Servings: One

Ingredients:

- Lemon Juice (1 t.)
- Maple Syrup (.50 T.)
- Rice Protein Powder (2 t.)
- Frozen Banana (1)
- Ice Cubes (6-10)
- Blueberries (20)
- Vanilla Soy Ice cream (.25 C.)
- Low FODMAP Milk (.50 C.)

Instructions:

1. Place all of the ingredients from above into a blender. Be sure to cut the frozen banana into smaller pieces.
2. Serve right away for a delicious breakfast.

Banana and Oats **FODMAP** Breakfast Smoothie

Prep Time: Five Minutes

Servings: One

Ingredients:

- Almond Milk (.50 C.)
- Linseeds (1 t.)
- Rolled Oats (1 T.)
- Banana (1)

Instructions:

1. Place all of the ingredients from above into a blender. Be sure you cut the banana into smaller pieces for easier blending.
2. Serve immediately for a filling and healthy meal.

Blueberry, Banana, and Peanut Butter Breakfast Smoothie

Prep Time: Five Minutes

Servings: One

Ingredients:

- Ice Cubes (6-10)
- Low FODMAP Milk (.75 C.)
- Blueberries (.50 C.)
- Peanut Butter (1 T.)
- Banana (.50)

Instructions:

1. Place all of the ingredients from above into a blender and blend until everything is smooth.
2. Serve immediately and enjoy!

Kale, Ginger, and Pineapple Breakfast Smoothie

Prep Time: Five Minutes

Servings: One

Ingredients:

- Ice (1 C.)
- Ginger (.25 T.)
- Kale (1 C.)
- Pineapple (.75 C.)
- Orange (.50)
- Low FODMAP Milk (1 C.)

Instructions:

1. Place all of the ingredients from above into a blender and blend until everything becomes smooth.
2. Serve immediately for a nice, healthy breakfast.

Strawberry and Banana Breakfast Smoothie

Prep: Five Minutes

Servings: One

Ingredients:

- Ice (1 C.)
- Maple Syrup (1 t.)
- Low FODMAP Milk (.75 C.)
- Strawberries (6)
- Banana (1)

Instructions:

1. Toss all of the ingredients into your blender and mix together until smooth.
2. Serve and for an extra treat, try adding some whipped cream!

Low FODMAP Soups and Salads:

Apple, Carrot, and Kale Salad

Prep Time: Ten Minute

Servings: Eight

Portion: .50 C.

Ingredients:

- Salt and Pepper (.25 t.)
- Maple Syrup (1.50 t.)
- Mustard (1 T.)
- Red Wine Vinegar (1.50 T.)
- Olive Oil (3 T.)
- Kale (.50 C.)
- Carrots (3)
- Apple (1 C.)

Instructions:

1. First step, you will want to create your dressing for the salad. You can do this by taking a small bowl and mixing together the maple syrup, mustard, vinegar, and oil. For some extra flavor, season with salt and pepper to taste.
2. Once this is done, take the kale, carrots, and apple and chop into fine, smaller pieces.
3. Finally, dress the salad, toss it a bit, and your meal is ready to be served!

Green Bean, Tomato, and Chicken Salad

Prep Time: Fifteen Minutes

Servings: Four

Portion: .50 C.

Ingredients:

- Lettuce (1 C.)
- Basil Leaves (2 T.)
- Cherry Tomatoes (10)
- Gruyere Cheese (.50 C.)
- Cooked Chicken (1 Lb.)
- Green Beans (.50 C.)

Instructions:

1. To begin, you will want to bring a medium pot of water to a boil. Once the water is boiling, cook your green beans for a few minutes. Once they are tender, drain the water from the pot and run the beans under cold water for a minute.
2. Next, take a large bowl and mix together all of the ingredients from above for a healthy salad.
3. For extra flavor, top your salad with any low FODMAP approved dressings.

Tuna Salad Low FODMAP Style

Prep Time: Five Minutes

Servings: Six

Portion: .50 C.

Ingredients:

- Salt and Pepper (to taste)
- Dried Dill (.50 t.)
- Lemon Juice (1.50 T.)
- Mayonnaise (.50 C.)
- Celery (.50)
- Tuna (2 Cans)

Instructions:

1. Start out by squeezing the liquid out of the tuna.
2. Once you have discarded the tuna, add it into a medium bowl with the vegetables from above.
3. When everything is stirred together, add in the dill, lemon juice, mayonnaise, along with the salt and pepper.
4. This mixture is great for any salad or sandwich!

Low FODMAP Pumpkin Soup

Prep Time: Ten Minutes

Cook Time: Fifteen Minutes

Servings: Six

Portion: 1 C.

Ingredients:

- Lactose-free Half and Half (.75 C.)
- Light Brown Sugar (1 T.)
- Canned Pure Pumpkin (1)
- Vegetable Soup Base (2 T.)
- Water (3 C.)
- Cayenne Pepper (.125 t.)
- Nutmeg (.25 t.)
- Cinnamon (.25 t.)
- Smoked Paprika (.25 t.)
- Scallions (.75 C.)
- Olive Oil (1 T.)
- Unsalted Butter (2 T.)
- Salt and Pepper to taste

Instructions:

1. To begin, you will want to heat up a medium sized pot over a low

to medium heat. Once the pot is warm, you can add in your oil and butter until it begins to sizzle.

2. When the butter and oil are warm, add in your spices with the scallions and cook until they are soft.

3. Once this happens, add in the soup and water. Be sure to mix everything together before you add in the salt, brown sugar, and the canned pumpkin.

4. Now that these ingredients are placed in the pot, lower your heat and allow these to simmer for ten minutes or so. Feel free to stir every once in a while, to assure the ingredients are blended well.

5. Now, remove the soup from the heat and add in your half and half. Once the soup is cool, you can place the mixture into a blender and blend until it is smooth.

6. For extra flavor, season the soup with salt and pepper to taste.

LOW FODMAP DIET

Quinoa and Turkey Meatball Soup

Prep Time: Fifteen Minutes

Cook Time: Twenty Minutes

Servings: Eight

Portion: 1 C.

Ingredients:

- Collard Greens (5 C.)
- Celery (.50)
- Leek Tips (1 C.)
- Olive Oil (2 T.)
- Egg (1)
- Dried Basil (2 T.)
- Parsley (2 T.)
- Cooked Quinoa (.50 C.)
- Ground Turkey (1 Lb.)
- Turkey Stock (10 C.)
- Salt and Pepper to taste

Instructions:

1. To start out, you are going to want to make your meatballs for the soup. You will do this by taking a large mixing bowl and combine the egg, parsley, basil, quinoa, and turkey together.

64

Gently take the mixture in your hands and form one inch balls.

2. Next, take a medium pan over medium heat and cook the turkey meatballs in olive oil for a few minutes. Be sure to flip the balls over so that they are a nice golden-brown color all around.

3. Now that these are done, take a large pot over medium heat and add in a tablespoon of oil. Once the oil is sizzling, you can add in the leek and celery. Sauté these two ingredients for a minute before adding in the collard greens and stock.

4. When all of the ingredients are cooked, add in the meatballs and allow this mixture to simmer over a low heat for eight to ten minutes.

5. Remove the soup from the heat and allow to cool slightly before serving.

Mixed Vegetable, Bean and Pasta Soup

Prep Time: Fifteen Minutes

Cook Time: Thirty Minutes

Servings: Fourteen

Portion: .75 C.

Ingredients:

- Gluten-free Pasta (1 C.)
- Dried Thyme (1 t.)
- Smoked Paprika (1 t.)
- Dried Basil (1 t.)
- Zucchini (1)
- Squash (1)
- Bok Choy (2 C.)
- Carrots (3)
- Kale (1 C.)
- Red Potatoes (1 C.)
- Butternut Squash (1 C.)
- Crushed Tomatoes (1 Can)
- Water (8 C.)
- Leek Tips (.25 C.)
- Scallions (.75 C.)
- Olive Oil (2 T.)
- Salt and Pepper to taste

Instructions:

1. To start, you will want to take a large pot and begin to heat it over medium heat with the olive oil placed in the bottom.

2. Once the olive oil is sizzling, add in the leeks and scallions and allow them to cook until they become soft.

3. When these are ready, add in your prepared zucchini, squash, Bok choy, carrots, kale, potatoes, chickpeas, canned tomatoes, and the water. Season as desired and place the top on the pot.

4. Bring all of the ingredients from above to a boil and then turn the heat down to allow everything to simmer for at least thirty minutes. By the end, all of the vegetables should be tender.

5. While the soup cooks, you can cook the gluten-free pasta in another pot so by the end, you can combine everything and have a healthy meal!

Vegan Options:

Low FODMAP Coconut and Banana Breakfast Cookie

Prep Time: Ten Minutes

Cook Time: Twenty Minutes

Servings: Ten

Portion: One

Ingredients:

- Vanilla Extract (1 t.)
- Vegetable Oil (.25 C.)
- Maple Syrup (.25 C.)
- Banana (1)
- Baking Powder (.50 t.)
- Cinnamon (1 t.)
- Ground Flax Seeds (2 T.)
- Chia Seeds (2 T.)
- Unsweetened Coconut Flakes (.50 C.)
- Banana Chips (.50 C.)
- Gluten-free All-purpose Flour (.50 C.)
- Old-fashioned Oats (1 C.)

Instructions:

1. You will want to begin by heating your oven to 325 degrees.

2. While the oven heats up, take a medium bowl and mix together the baking powder, cinnamon, flax seeds, chia seeds, coconut flakes, banana chips, flour, and oats altogether.

3. In another bowl, mix together a mashed banana, vanilla, vegetable oil, and pale syrup. When both bowls are well combined, you can mix them together and begin to create your dough.

4. Next, take a greased cookie sheet and lay out balls of dough to create your cookies. When this is done, pop the cookie sheet in the oven for twenty minutes.

5. When the time is up, remove the cookies, allow to cool, and enjoy!

Lemon and Garlic Roasted Zucchini

Prep Time: Five Minutes

Cook Time: Twenty Minutes

Servings: Twelve

Portion: 1 C.

Ingredients:

- Olive Oil (1.50 T.)
- Zucchini (2)
- Lemon Zest (2 T.)
- Salt and Pepper to taste

Instructions:

1. You can begin by heating your oven to 425 degrees.
2. While this warms up, slice your zucchini into thin slices and place in a bowl with the lemon zest and olive oil. Assure it is covered completely before seasoning with salt and pepper.
3. Place the zucchini on a greased sheet pan and cook for twenty minutes.

Rainbow Low FODMAP Slaw

Prep Time: Ten Minutes

Servings: Twenty

Portion: 1 C.

Ingredients:

- Pomegranate Seeds (.50 C.)
- Carrots (3)
- Kale (1 C.)
- Red Cabbage (1 C.)
- Green Cabbage (1 C.)
- Lactose-free Yogurt (.50 C.)
- Dijon Mustard (1 t.)
- Sugar (2 T.)
- Apple Cider Vinegar (.25 C.)
- Canola Oil (.50 C.)

Ingredients:

1. Start out by creating your dressing for the slaw. You can do this in a small bowl, mix together the canola oil, apple cider vinegar, Dijon mustard, sugar, yogurt, and a little bit of salt.
2. In another bowl, toss together the different cabbage with the carrots and the kale.
3. Gently drizzle the dressing over the kale, and you have a delicious slaw that is full of color and flavor!

Vegan Roasted Red Pepper Farfalle

Prep Time: Ten Minutes

Cook Time: Ten Minutes

Servings: Four

Portion: 1 C.

Ingredients:

- Capers (.25 C.)
- Parsley (.75 C.)
- Olive Oil (.25 C.)
- Roasted Red Peppers (1 Jar)
- Gluten-free Farfalle Pasta (2 C.)

Instructions:

1. You can start this recipe by cooking your pasta according to the instructions on the side of the box.
2. Once the pasta is cooked through, drain the water and then place the pasta back into the pot.
3. Toss in the oil, parsley, roasted red peppers, and capers to the mixture.
4. Mix everything together and season with salt and pepper for extra flavor.

As you can tell, you can follow the low FODMAP diet and still enjoy delicious foods. While these are only some of the many recipes you can follow on your diet, there are plenty of resources out there to provide you with even more! With these resources in hand, we will now go over a simple seven and fourteen-day meal plan that is easy to follow.

With a limited food choices, you may be thinking to yourself that you are going to get bored quick. When it comes to a new diet, it is all about your frame of mind. On one hand, you could think negatively about it and return to your old eating habits. With choice, comes consequence. When you eat the foods that trigger you, you are going to feel lousy. Why make that choice when you can choose to eat healthy and feel better? Below, we will provide some simple meals for you to consider until you feel confident enough to create your own recipes

Breakfast Meal Plan Ideas:

- Eggs- Hard-boiled, over easy, or even scrambled. There are many ways to enjoy eggs!
- Lactose-Free Yogurt with any low FODMAP fruit
- Gluten-free Muffins
- Gluten-free French toast
- Gluten-free Oatmeal with cinnamon
- Rice Cereal with low FODMAP fruit
- Ground Turkey
- Smoothie with low FODMAP fruit

Lunch Meal Plan Ideas:

- Gluten-free Bread with Deli Meat and Cheese
- Chicken Noodle Soup
- Quinoa Bowl with low FODMAP Veggies or Grilled Chicken
- Salad
- Baked Potato with Lactose-free Butter

Dinner Meal Plan Ideas:

- Stir-Fried Rice
- Tacos
- Gluten-Free Pizza
- Grilled Chicken Salad
- Steak with Fresh Low FODMAP Vegetables
- Grilled Chicken with White Rice
- Rice Pasta with Marinara
- Snack Meal Plan Ideas:
- Rice Cakes with Peanut Butter
- Baby Carrots
- Lactose-free Yogurt
- Unripe Banana
- Unsalted Peanuts
- Pop Chips
- Gluten-free Pretzels
- Crackers with Cheese
- Hard-Boiled Egg

14- Day Meal Plan

Week One:

Meal	Monday	Tues.	Wed.	Thurs.	Friday
BFast	Small Banana Pancakes	Blueberry Smoothie	Roasted Sausage and Vegetable Breakfast Casserole	Strawberry and Banana Breakfast Smoothie	Banana and Oats FODMAP Breakfast Smoothie
Lunch	Apple, Carrot, and Kale Salad	Mixed Vegetable, Bean, and Pasta Soup	Low FODMAP Pumpkin Soup	Tuna Salad Low FODMAP Style	Quinoa and Turkey Meatball Soup
Dinner	Low FODMAP Veggie Latkes	Steak with Lemon and Garlic Roasted Zucchini	Left Over Mixed Vegetable, Bean, and Pasta Soup	Vegan Roasted Red Pepper Farfalle	Salad with Grilled Chicken and Homemade Dressing

Meal	Saturday	Sunday
Breakfast	Eggs and low FODMAP fruit	Rice Cereal with low FODMAP fruit
Lunch	Chicken Noodle Soup	Baked Potato with Lactose-free Butter
Dinner	Stir-Fried Rice	Gluten-free Pizza

Week Two:

Meal	Monday	Tuesday	Wednesday	Thursday	Friday
Breakfast	Gluten-free French Toast	Rice Cereal with low FODMAP fruit	Lactose-free Yogurt with low FODMAP fruit	Blueberry Smoothie	Small Banana Pancakes
Lunch	Quinoa Bowl	Salad with Approved Dressing	Gluten-free Sandwich with Deli Meat and Cheese	Mixed Vegetable, Bean, and Pasta Soup	Tuna Salad on Gluten-free Bread
Dinner	Grilled Chicken with White Rice	Grilled Chicken Salad	Salad with Approved Dressing	Gluten-free Tacos	Stir-Fried Rice

Meal	Saturday	Sunday
Breakfast	Smoothie with low FODMAP fruit	Lactose-free Yogurt with low FODMAP fruit
Lunch	Rice Pasta with Marinara	Chicken Noodle Soup
Dinner	Baked Potato with Lactose-free Butter	Grilled Chicken Salad

Vegan 7-Day Meal Plan

Meal	Monday	Tuesday	Wednesday	Thursday	Friday
Breakfast	Coconut Yogurt with Chia Seeds	Rice Cakes with Peanut Butter	Corn Flakes with Almond Milk	Gluten-free Bread with Almond Butter	Unripe Banana with Coconut Yogurt
Lunch	Lemon and Garlic Roasted Zucchini	Rainbow Low FODMAP Slaw with Gluten-free Bread	Vegan Roasted Red Pepper Farfalle	Low FODMAP Coconut and Banana Cookie with Coconut Yogurt	Salad with Approved Dressing
Dinner	Low FODMAP Veggie Latkes	Gluten-free Pasta with Approved Sauce	Plain Tempeh with low FODMAP Veggie of choice	Plain Tofu with Rice Noodles	Gluten-free Pizza with Soy Cheese

Meal	Saturday	Sunday
Breakfast	Blueberry Smoothie with Coconut Milk	Banana and Oat Smoothie with Coconut Milk
Lunch	Plain Tofu with Soba Noodles	Plain Tempeh with Gluten-free Pasta
Dinner	Grilled Cabbage Soup	Baked Brussel Sprouts with Plain Tofu

Chapter 6: Low FODMAP diet tips and tricks for success

Starting a new diet can be scary. As we said before, your frame of mind is going to be incredibly important. It is vital you think about your why when making food choices. Each meal, we have a chance to better our health; all it takes is a little thought behind each decision.

Of course, we want to see you succeed with your diet. Below, you will find a number of tips and tricks that have helped other clients on the low FODMAP diet. While some may work for you, others may not. You must adjust the low FODMAP diet to match your desired lifestyle so you not only stick with it but can enjoy it at the same time!

1. Read the Label

Reading the labels on packaged foods is going to be vital for the success of your diet. Unfortunately, many high FODMAP ingredients can have very confusing names. We suggest carrying a list of additives to avoid until you learn them by heart. When you are more aware, you can avoid the high FODMAP ingredients.

2. Water-soluble

In general, low FODMAP foods are going to be water-soluble, but this does not mean they are fat-soluble. If you are cooking a soup with an onion, you will want to take the onion out. Instead, try using onion-infused oils for the taste. It is quick fix that may help with your IBS triggers.

1. High Fructose Corn Syrup

High Fructose Corn Syrup is in everything. Again, it will be important that you learn how to read food labels so you will be able to avoid this mistake. This ingredient is in a number of foods including energy bars, juices, mayonnaise, frozen meals, and even popcorn. Check the label before you put anything into your shopping cart.

2. Fiber

If you pick up a product and it seems to have a high serving f fiber, you can assume it is due to a high FODMAP additive. Try to avoid any products that boast about their fiber; it's a trap! Any fiber additives will more than likely trigger your GI issues.

3. Onion and Garlic Powder

When it comes to choosing out your spices, pay special attention to the labels. You will want to avoid onion and garlic powders as they contain high FODMAPs. Luckily, there are plenty of delicious low FODMAP approved spices as you can find in the chapter from above.

4. "Natural"

If you find any frozen foods, brothers, or savory soups that claim they have "natural" flavors, go ahead and check out the label. You can assume that they contain garlic and onion, very popular IBS triggers. You will want to try your best to avoid these additives to your meals.

5. "Healthy"

As much as we would like to trust when products claim they are healthy, this does not equate to low FODMAP approved. Foods like asparagus and apples are supposedly "healthy" for you, but they can trigger IBS symptoms. As you go through the elimination process, you will learn just what you can and cannot eat and make the decision if something is healthy for you.

6. Beverages

Often times, people forget that beverages can contain FODMAPs. You will want to pay special attention to what you are putting into your body. If you ever have questions, feel free to refer to our lists in the chapter from above. Just because a beverage claims it has no net carbohydrates, this does not mean they aren't high in FODMAPs.

7. Portion Control

While you are on the low FODMAP diet, portion is going to be key to success. When you are reading labels, you will always want to pay special attention to portion size. While a low FODMAP diet is approved, a bigger portion may still trigger IBS symptoms. You will want to try your best to be mindful of portion control.

8. Learn About Yourself

As you start this diet, you will want to spend plenty of time on the elimination phase. The more you test, the more you will be able to figure out what foods you can and cannot eat. When you have more to choose from, you will be able to get more creative with your recipes. At the end of the day, only you know what is best for you. When you learn yourself, the diet will become that much easier.

9. Food and Meal Journal

Your food and meal journal are going to be an important tool for your low FODMAP diet journey. By keeping track of the foods, you can and cannot eat; it will make it easier when you go back to check out your history. We eat so many different types of foods through the day; it can be hard to remember which foods trigger you. By keeping a journal, it leaves little room for mistakes.

10. Use Your Fridge

If you are trying your best to stick with the low FODMAP diet, why leave anything up to chance? Do yourself a favor and take the time to remove any high FODMAP foods in your house. By keeping your pantry and fridge stocked with the low FODMAP foods you need, it deletes any temptations you may have in the house.

11. Have A Backup Plan

Dieting is hard, especially when you are first starting out. When you are planning out your meals, it is possible to miss one here or there. Try to stack your freezer with low FODMAP meals so you can cook them in a few minutes. When your first plan falls through, you will always have a backup. It is a win-win situation!

Chapter 7: Low FODMAP diet FAQ

As we are nearing our time together, hopefully, you are feeling better about starting the low FODMAP diet. While you have learned a lot about the diet, feel free to check back whenever you have a question about the diet. Whether you need a refresher on the benefits of the diet or a reminder of which foods you can and cannot eat, you will be able to find the information here easily.

To finish off, we will hopefully be able to answer any further questions you have about the low FODMAP diet. Simply remember that this diet is going to be specifically tailored up to you. Being diagnosed with IBS or other GI tract issues is not the end of the world. It will take some extra effort, but when your symptoms and discomfort are relieved, you will be thankful you made the choice to start the low FODMAP diet. For now, it is time to answer some more popular questions you may still have.

Q. I am following the low FODMAP diet and still experiencing symptoms, is this the right diet for me?

A. The answer could be yes and no. If you are following the diet and still find yourself with symptoms of IBS, there may be another culprit in your diet. Remember to keep a food diet with you at all times so you can find any triggers you may be missing.

Q. Can I follow the low FODMAP diet as a vegetarian?

A. Absolutely! You can follow this diet whether you are vegan or vegetarian, it will just take a little extra work. You will find in the chapters before there are plenty of choices, so long as the allowed foods are not triggers for your own body. Some good sources of protein for this diet would be chickpeas, tofu, tempeh, and more. If there is a will, there is always a way!

Q. How do I make sure I'm getting enough Fiber?

A. This is one of the bigger concerns for those following the low FODMAP diet, especially if constipation is an issue. Luckily, several good low FODMAP sources can help you keep your fiber intake up. These include chia seeds, brown rice, flax seeds, kiwi, oranges, white potato, rice bran and more. Check out the list provided in the fourth chapter for a longer list.

Q. Should I eat Larger or Smaller Meals on the Low FODMAP Diet?

A. In general, you should try to eat three main meals through your day and to snacks between these meals. If you are still hungry, you can always add in another snack. Remember that portion control is going to be vital while you are on the low FODMAP diet, so this is something you will want to keep in mind when plating your meals.

Q. What is the rule with fats and oils on the Low FODMAP Diet?

A. As a general rule, there are plenty of fats and oils that are low in FODMAPs. However, anything in excess can trigger IBS symptoms. You will want to be especially aware of any condiments or sauces that are oil-based such as salad dressings. Most of the time, these also include high FODMAPs like garlic. Remember to always read the labels before consuming anything. You can also refer to our extended grocery list to see which fats and oils are allowed on the diet.

Q. Can I eat meat on the low FODMAP Diet?

A. Yes and No. Some sources of animal protein such as fish and chicken are low in FODMAPs. However, if the meat is prepared already, you will want to avoid any additives that may trigger your symptoms. If you have any further questions, please refer to the food lists from the chapter above.

Q. What happens if I break my diet?

A. While the aim of the diet is to stick to it as much as possible, mistakes and slip ups will happen. Overall, you will want to achieve control over any symptoms you may be having. If you slip up, expect to experience the IBS symptoms. As long as you return to your diet, you will most likely be able to improve them in a few days.

Q. Is this a lifestyle?

A. No, the low FODMAP diet is not meant to last for a lifetime. The aim of the diet is to help heal your gut over a controlled period of time. This diet should only be followed for two to six weeks. After this, you can begin to introduce food back into your diet. This will change depending on each individual.

PART 2

The following chapters will discuss all of the many health benefits to maintaining an alkaline diet as well as the keys to unlocking all of your health potentials. We will look at the chemistry behind the pH scale and how it works and behave within our bodies. We will learn all about why an alkaline balanced diet is the best way to keep your body as healthy as possible.

Maintaining a happy and healthy body can be a real challenge, but within the pages of this book, you will find all of the secrets and tips for making an alkaline-based pH balanced diet a major part of your life with high-impact solutions and surefire methods for creating a healthy pH balance in your own body chemistry.

We will also look at some amazing meal ideas to keep you going strong on the alkaline diet all day every day. Don't miss a single meal with these new and exciting meal ideas. Never be in the dark again! After reading this book, you will know everything you need in order to create and maintain a healthy and satisfying alkaline diet.

There are plenty of books on this subject on the market, so thanks again for choosing this one! Every effort was made to ensure it is full of as much useful information as possible. Please enjoy!

Chapter 1: What is the Alkaline Diet?

We eat the foods that we eat for all kinds of different reasons. Sure, from an evolutionary standpoint, we eat food so that we can take in calories and convert them into energy in order to fuel our bodies and keep us going throughout the entire day. The food we eat also provides us with the essential nutrients that our bodies need in order to keep them running in an optimal manner.

But we also eat food for pleasure, the sheer joy of tasting something amazing that we truly love. We eat socially. Food has been a way of bringing people together since the very dawn of civilization. Sometimes we use food for comfort and sometimes we use it to mark formal and important occasions. We use food as a proving ground over which to test out new prospective romantic partners.

And yet, for all that food can do for us, so many of us take it for granted and don't seek out ways to make our food work for us. Used correctly, food and nutrition are tools that can turn our bodies into the healthiest and efficient powerhouse that nature intended them to be.

With so many different diets and nutrition plans out there, it can be hard to know which one is right for you. Well, you're reading this book so you already know that you're on the right track!

Indeed, the alkaline diet is a tried and tested way to get the most out of your body. But how does it work? Why does it work? How can eating an alkaline diet optimize your body and health?

The key to understanding the science and chemistry behind how the foods we eat affect us is to understand that just like the fundamental laws of physics, every action has an equal and opposite reaction. Or in other words, everything that we put into our bodies will affect us based on the characteristics of that particular food item. So if we eat a lot of things that cause the same or similar effects on our bodies, we can influence and even control the effects and changes that our body takes by carefully selecting the foods that we eat and what effect they have on our bodies. As the popular saying goes — you are what you eat.

To illustrate, imagine you are walking through the woods and you accidentally brush up against a poison ivy leaf. Well, sorry to say it, but there is a very good chance that you are going to develop an itchy poison ivy rash. If however, you get completely naked and roll around in an entire patch of poison ivy, you are pretty much guaranteed to get poison ivy and a whole lot of it!

Humorous examples aside, it stands to reason that if we know that a particular food or food group has a particular effect on our body, we can effectively control any number of internal body systems by carefully planning and selecting the foods we eat.

So how does the alkaline diet promote health? Well, the alkaline diet is all about balance. So many of the negative health issues in our lives are the result of an imbalance in our bodies. So much of the history of medicine revolves around finding the ideal balances for the human body.

For many, many years, doctors around the world attributed all of our health conditions, whether good or ill, to a balance or imbalance in what they referred to as the "humors". As far back as Ancient Greece and Ancient Rome, there was a near-universal belief that four humors or bodily fluids influenced every aspect of health and temperament, and ill health or ill temperament was the result of deficiencies or excesses on one or more of these four humors. These four humors were black bile, yellow bile, phlegm, and blood. Each of these four humors was associated with a particular personality type and other such characteristics.

When a person came to an ancient doctor with an ailment, the ancient doctor would examine their patient to determine their temperament and along with other factors would determine where their imbalance in humors was, and then they would come up with a treatment plan with the intention of balancing the patient's humors. So in other words, for millennia, the goal of medicine has been to achieve balance in the human body.

And while many of the theories and practices of ancient physicians have long ago fallen out of use in favor of new techniques and schools of thought, modern science has nevertheless confirmed at least some aspects of ancient medicine, namely, the concept of balance itself.

While we don't hear much about black bile, yellow bile, or phlegm anymore in modern medicine, the fourth humor that ancient doctors treated is certainly still extremely prominent in modern medicine — blood.

Blood is still very much our life force just as it was believed by ancient doctors. Blood is the fluid that keeps us living and breathing and a proper medical understanding is absolutely integral to maintain overall good health.

So how can we maintain a good balance in our blood? What aspect of our blood do we even need to balance? What negative effects can we avoid by maintaining balanced blood and what positive ones can we promote?

While those ancient doctors were certainly on the right track, they had a few key factors wrong so, in order to move forward, we are going to need a firmer and more modern grasp of the science behind our health and nutrition.

To understand this concept a little bit better, we need to understand one of the most fundamental aspects of chemistry. This integral part of chemistry and science as a whole is known as the pH balance or the pH scale. We are going to learn all about the pH balance or the pH scale and how it can affect our bodies in a positive way in the following chapters. First, we will learn what a pH balance is.

Chapter 2: What is a pH Balance?

The first thing we need to understand on our journey to the perfect internal balance via the alkaline diet is exactly just what the pH balance is. Furthermore, we need to understand how the chemical characteristics of a substance or fluid play a role in determining where it falls on the pH scale.

What exactly does the pH scale measure? In short, the pH scale is a measure of the acidity or basicity of solution in which the solvent is water. Such a solution is known as an aqueous solution. In other words, when a substance is dissolved in or is otherwise mixed with water, it can then be tested and measured on the pH scale.

An aqueous solution can be defined as either an acid or a base, as this is precisely what the pH scale is meant to determine. An aqueous solution that is basic is referred to as being alkaline. This gives us a pretty good indication of what the alkaline diet is all about. The pH scale itself is a type of scale known as a logarithmic scale. This means that each equidistant quantified measure is an order of magnitude greater than the previous measurement on the scale. The scale ranges from zero to fourteen, with a neutral pH value being in the middle at seven.

Solutions that have a pH value of less the median value of seven are defined as being acidic, while the opposite scenario, in which a solution is measured to have a pH value of higher than seven — that solution is called basic. Water that is pure and unadulterated is pH neutral which is to say that it should prove to have a pH value of seven when tested, as natural dihydrogen oxide, the chemical name for water is neither a base nor an acid. If that is not the case, then the water should be tested for impurities.

While it is possible for an aqueous solution to have a pH value greater than fourteen or less than zero, these would have to be extremely acidic or extremely basic solutions and would not only be decisively deadly to ingest and even extremely dangerous just to touch. Therefore, for practical purposes, official pH values are nearly always represented on a scale between zero and fourteen.

The pH scale is defined by a set of international standards that are determined and agreed upon by an international scientific body. There are several ways to test the pH level of an aqueous solution, with one of the most notable ones being the use of a glass electrode combined with a pH meter. This scientific instrument determines the difference between a pH electrode and a control electrode in terms of their respective electrical potential. This difference in the electrical potential of a solution relates directly to the acidity of the solution and can therefore be used to give it a standard value.

Another very popular and frequently used means by which to test the pH value of an aqueous solution is by using one of the various compounds known as pH indicators.

A pH indicator is generally some kind of substance that when mixed with an aqueous solution results in a chemical reaction that will literally change the color of the solution, and by examining and comparing the color of the resulting solution, the pH value of the solution can be determined.

There are other pH indicators that indicate the pH level of a solution by chemical reactions that result in other physical indicators such as odor for example. However, by far the most common variety of pH indicators are visual in nature, generally based around color.

One of the most common types of pH indicators is the naturally occurring family of chemical compounds called anthocyanin. These compounds naturally change color reflects the pH balance of whatever item the compound is found within. Generally, these compounds are found in colored plant leaves or other plant parts. One of the most common sources of these pH indicating compounds is from the leaves of a red cabbage. The reason for this is because it is quite easy to extract anthocyanin from a red cabbage making it the perfect resource for a homemade pH indicator test for either health or educational purposes.

Anthocyanin can be found in many different plants though, such as the leaves of the aforementioned red cabbage, but also in certain flowers such as the geranium, the poppy, and also rose petals. Berries and stems can also house anthocyanin compounds such as blueberries and blackcurrants as well as rhubarb. In short, most plants or vegetables that have reddish, purplish, or bluish color in all likelihood contain at least a small amount of anthocyanin compound. When used as a pH indicator by mixing it with an aqueous solution, an anthocyanin compound will become redder the more acidic the solution is and will turn from red to purple to blue the more alkaline the solution is.

Another very commonly used pH indicator since medieval times is the substance called litmus which is derived from various species of lichen.

In fact, the word litmus itself means colored moss in its original language, Old Norse. Just like anthocyanin compounds, litmus will turn red when exposed to acidic solutions and blue when exposed to basic solutions. You may even be familiar with the term 'litmus test'. It has come to be used very commonly and very broadly as a metaphor for anything that could be used to soundly distinguish between multiple options.

So with pH balance being fundamental to the chemical nature of all kinds of biological material including the foods we eat, how do we know if and how such foods are affecting our health? We will continue learning about pH imbalance in our bodies to find out. The next chapter will go into the science of how pH balance or more specifically imbalance can affect our bodies and our health.

Chapter 3: The Science Behind pH Imbalance

Every single substance in the world has a pH balance and that includes all of us. Sure, we don't make cabbages change color when we pick them up, but our bodies must maintain a certain pH level in order to live and function properly. This pH balance that is naturally maintained in our bodies is called the acid-based balance and it is quite literally exactly what it sounds like — the balance of acidic and basic substances in your body. More specifically though, when we are referring to the acid-base balance of our bodies, we are most often referring to the pH balance of our blood.

The human body is designed with a few systems in place intended to keep the natural pH levels regulated at an appropriate balance between acidity and alkalinity. Both the kidneys as well as the lungs have a very important role to play in this process. As we previously laid out, the pH balance is generally expressed as a value between zero and fourteen, with seven being the neutral value. And remember that pure and unadulterated water should have a pH value of exactly seven. Knowing then that water is neutral seven on the pH scale, and knowing also that our bodies are designed to maintain an even pH balance, it would stand to reason that our blood should have a neutral pH value of seven as well, right?

Well, not quite. And this is a major key to understand the alkaline diet. The ideal blood pH level is not actually a neutral seven but instead generally should be about a 7.40 on the pH scale. This value can vary slightly from person to person, but that is the standard average. And yes, that's right— the human body should have a blood pH level that is a little bit on the alkaline side.

Generally speaking, it is the kidneys and lungs that regulate this pH level, so if they are not functioning normally, the blood pH level can become imbalanced. This kind of pH imbalance can lead to serious medical conditions which are called acidosis or alkalosis depending on which direction the imbalance goes in. It is important to note that these serious medical conditions must be treated by a medical professional and diet alone cannot entirely reverse these conditions.

Now, what we're talking about in this book is the small, minor imbalances that a general practitioner wouldn't be concerned about because they aren't severe enough to have a serious debilitating effect, but that certainly do have your body operating in sub-optimal conditions, and more importantly, the alkaline diet that can have it function far better than you ever imagined possible.

So in order to better understand how the alkaline diet will allow us to correct these small but important pH imbalances, we'll need to have a complete understanding of what could throw our pH out of balance and why it might happen.

As we have established a moment ago, the primary regulators of the body's pH level are the kidneys and the lungs. There are a large number of small systems in our bodies that have their own pH level and regulate them in their own ways, but the two main, body-wide regulators are the lungs and kidneys.

As you are likely already aware, we take in oxygen with our lungs when we inhale and expel carbon dioxide when we exhale. The oxygen that we take in is absorbed inside our lungs and used as fuel by our cells. The waste product that our cells produce by using the oxygen is carbon dioxide. Which is all very simple and pretty straightforward and familiar to all of us but here's the important part — carbon dioxide is slightly acidic. So by making slight adjustments to how much carbon dioxide is released or retained, our lungs are able to make adjustments to the overall acid-base level of our blood.

Similarly, the kidneys being the filtration system for the vascular system have the ability to excrete small amounts of acidic or basic compounds into the blood in order to make slight alterations to our blood chemistry. This is a slow process as compared to the more immediate effect of the lungs' pH regulatory system. So the lungs and the kidneys could be thought of as the short-term and long-term blood pH level regulators of our body.

If the blood pH level is out of balance, then it can lead to one of these two conditions — alkalosis and acidosis. With the standard balanced blood pH level being 7.40, anything below 7.35 is considered acidosis and anything above 7.45 is called alkalosis. Again, it is important to note that these are serious medical conditions and must be treated by a medical professional. It is always best to consult your doctor if you are suffering from these conditions. What we can do, however, is assist our body's natural pH regulation system by maintaining a blood pH level that is within the tolerable limits.

A low blood pH level or in other words, slightly acidic blood is far more common than the inverse and so that it is what we are primarily focusing on — an alkaline diet that will help us maintain a healthy blood pH level.

While any level measured at 7.35 and under is acidosis and needs professional medical treatment, it is far too common for our blood pH level to fall into that 'safe' range of 7.36-7.39 without being at that ideal sweet spot of 7.40. If you want to get the most out of your body, if you want your body to be operating at peak performance, and if you want to live your absolute healthiest life, then the 'safe' level of 7.36 is not tolerable for your body.

If you are truly serious about your health and your wellbeing, then the 'safe' blood pH level of 7.39 isn't even good enough for you. You need to have the absolute optimal blood pH level and you will settle for nothing but a perfect 7.40. Continue on reading in the next chapters and we are going to show you how.

Chapter 4: Why Alkaline is Best

If our body's pH level is all about balance, then why would maintaining an alkaline diet be superior to an acidic one? Shouldn't we be consuming a perfect balance of alkaline and acidic foods and nutrition? If those are among the questions you are now asking yourself, then you are on the right track. Those are great questions to ask.

There are several reasons why an alkaline diet is a crucial component in maintaining a healthy body and blood pH level. Remember that magic number? The ideal pH level for our blood that will allow our body to operate optimally? That is right — it was 7.40. And do you remember what the pH value for perfectly pure, unadulterated water is? That is right — it was a perfect seven. So what that means, of course, is that the ideal blood pH balance is in fact slightly alkaline at 0.4 units more basic than water.

So we can see already that in order to maintain our ideal pH balance, we will need to intake more alkaline foods than acidic foods. Of course, that is not to say that you can never consume anything acidic. In fact, it is important to have acids as well in order to maintain balance. We just need to be perfectly aware that our body does in fact require a slight alkaline balance and so we should be mindful of this when we plan our meals and overall diets.

This balance may also be reflected in the foods we choose to eat. They don't necessarily need to be extremely alkaline in order to transfer to us the health benefits we are looking for. They may only need to have a slight pull on the alkaline side of the scale. It all depends on our individual bodies and what they are in need of. And of course, everything scales. So a lot of a slightly alkaline substance may have the same value as a little of something with a higher alkaline value. Remember as well that the pH scale is logarithmic which is to say that each unit is exponential to the value of the previous unit. That means that consuming something with an alkaline value of nine would be ten times more alkaline than something with the alkaline value of eight. This is why we need to be careful when consuming anything that is alkaline or acidic. Things can become unhealthy or even dangerous in a real hurry. So, remember to plan ahead and do everything in moderation.

Another equally surface-level reason why it is important to consume a healthy amount of alkaline rich foods is because whether we are aware of it or not, many if not most of the foods we eat on a regular basis are either slightly or moderately acidic. Some very common foods and beverages even go as far as being highly acidic.

Now, again, it bears repeating that this does not mean that you cannot or should not consume these types of acidic foods and beverages at all. In fact, some of these acidic foods and beverages are very healthy and high in essential nutrients. The important thing, however, is to be aware of how much acidic substances we are consuming and how acidic those substances are.

Do you like fruit juices? How about coffee? Those are two great examples of highly acidic beverages that many of us consume on a regular basis. That is not necessarily a bad thing but just think about this — are you taking in the necessary amount of alkaline foods or liquids in order to maintain a healthy and optimal balance?

And what's more, it can often be a good deal more complicated than whether the particular food item that we are consuming is acidic or alkaline on a surface level. What makes the important difference is how the item we consume affects our blood pH level after it has been metabolized. And that could, in fact, be a good deal different than what it might seem to be based on the original acidity or alkalinity of the food item in question.

Another very important reason to remember to include alkaline foods in our diet in order to maintain a good acid-base blood balance is that an acid rich environment is considered by medical professionals to be a hotbed of disease and illness. And remember, it doesn't take much to become imbalanced in one's blood levels, so even a minor imbalance could quickly become a breeding ground for all manner of illness and health problems that you will absolutely want to avoid.

Just by remembering to consume a healthy and appropriate amount of alkaline foods and drinks, we can be safeguarding ourselves from any number of serious health concerns that could be lurking in our very blood. If you want to kill all of those potential illnesses dead before they become a real concern, you will need to act now and ensure that you are consuming an appropriate amount of alkaline foods.

This is the very topic that we will be going into next. We now know how important it is to maintain a good acid-base blood balance. We now know what that optimal blood pH balance is. And most importantly, we now know the dangers associated with having blood that is too acidic, and why it is so common for us to have a blood pH level that skews a little bit too acidic but not enough to go into full acidosis.

Equipped with this crucial information, we can now move on to learning about how to apply these factors to our everyday lives. Now, we are going to learn everything we need to know about how to create and maintain a balanced blood pH level, and all the tips and tricks to make it easy and straightforward.

Are you ready to have a body operating at optimal health? Are you ready to get the most out of your diet? Are you ready to prevent disease and illness that you didn't even know you were susceptible to?

Then continue on reading on, because all of your questions are about to be answered.

Chapter 5: Creating an Acid-Alkaline Balance

In this chapter, we are going to take a look at some of the biggest and best ways to gain control of your acid-alkaline blood levels and ensure that you can maintain them at an optimal level. Many of the things that we are going to talk about here are not just about diet. In fact, even the alkaline diet is not just about diet — it is about habits. It is about keeping good habits, maintaining regular health goals, and being in tune with your own body.

There are plenty of signs and symptoms that you may notice in the event that your body is too acidic. It is very likely that you will be experiencing chronic fatigue if your body is too acidic. Even if it seems that though you have been sleeping enough, you may still feel this way. Other symptoms of overly acidic blood are pain, headaches, joint pain, and stiffness.

Generally, people with acidic blood express an overall feeling of sluggishness and lethargy — sometimes even depression. It is also associated with a sense of irritability and a dulling of the mental faculties.

Obviously, if you are experiencing any of these symptoms, it will be in your best interest to correct them to the best of your ability. There are lots of ways to make your blood more alkaline and we will look at a few here.

First of all, you will want to make sure that your symptoms or feelings are in fact coming from a pH imbalance. In order to do that, you will need to check your blood pH levels regularly in order to maintain an up-to-date record of your pH levels. You can do this very easily with simple, inexpensive home testing kits available online and at many drug stores. You can get an instant and highly accurate reading and find out exactly where your body's pH balance is sitting. These simple test kits can help you make healthy and informed decisions about your personal health based on accurate and current information. This is a great, convenient, and inexpensive way to always be on top of your health.

Now, before we get into specific diet plans, let us talk about some changes that we can make to our diet in a general sense that will help improve our blood pH levels. One such thing that we can do to ensure that we are maintaining appropriate levels of acidity in our blood is by making sure that we eat more greens and dark-colored vegetables in general. Greens are not always the most popular foods to eat despite their great reputation and association with good health. But there are ways to make greens and other veggies fun and exciting and taste great.

You could try new recipes and try new types of veggies that you have never tried before. If you already love veggies, try to make sure that you get a good amount on a regular basis, if not every day. Even if you have a particular proclivity for veggies, it's easy to leave them out on occasion. Try to avoid that tendency.

And if you don't like veggies, maybe it has something to do with a reduction in their appeal on account of processed foods and excessive artificial sugars. Simply by cutting these things out of our diets as much as possible can dramatically reduce cravings for said items and make good, nutrient-rich foods like dark, green veggies far more appealing.

Either way, try to experiment with new ways of getting your veggies and making them fun and enjoyable. Try keeping some prepared in advance so that you always have a quick and healthy snack.

Here is another quick tip for general health and well-being. Every morning, the first thing you do when you wake up should be to drink a great big glass of ice-cold water as fast as you can. Why? Well, it's quick and it's easy, it costs nothing, and it has all kinds of great health and wellness benefits both short-term and long-term. First of all, the most immediate benefit is that ice-cold blast of invigorating water will snap you wide awake quicker and more effectively than any caffeinated beverage.

What's more, there's nothing better than an icy surprise to jump-start your body and kick your metabolism into a high-gear first thing in the morning. And because our friend water is completely calorie-free, that basically amounts to an energy boost and metabolism enhancer for free, metabolically speaking.

And do you want to take this brilliant life hack one step further and bring it into our alkaline friendly lifestyle? Add just a touch of lemon to that morning burst of water, or better yet, all the water you drink and you will get all the previous benefits plus that boost to your body's alkalinity that you need to function at peak efficiency. This may seem counter-intuitive given that lemon is acidic, but remember, it is not always about the acid-alkaline balance of the compound itself. It is how our bodies metabolize that compound. And lemon, being a well-known metabolism booster, will give you that alkaline push you need.

Of course, sometimes it is about the actual acidity of the food we are eating. Specifically, it is about the amount of acidic foods we are eating. If you find that you are suffering from symptoms of acid reflux, kidney stones, low bone density, or anything else associated with high body acidity levels — that is almost certainly a strong indication that you should be strictly limiting your intake of acidic foods.

This goes of course for any of the obvious culprits like tomato sauce or spicy foods, but there are some less obvious foods that metabolize into an acidic by-product in our bodies that we should be cautious of as well. This includes many processed cakes and cereals, often grain such as rice, oats or pasta, and even certain nuts like peanuts or walnuts. The key here is to just always be aware of what we are consuming and keep everything in moderation.

Beverages as well should be kept in moderation especially coffee and alcohol since both of which are associated with many negative health effects when consumed in excess, far beyond body acid-alkaline balance.

Chapter 6: Alkaline Diet for Vegetarians

Don't let the title fool you, this isn't just for vegetarians. The alkaline diet is great for anyone and everyone. If you are already a vegetarian — great, you are already in a great spot to maintain an amazing alkaline diet. If you're not a vegetarian, that's okay too. Remember in the previous chapter when we said that it is not necessarily about how acidic the foods we eat are but the quantities? Well, that goes for meat. Most meats are extremely acid forming in our bodies.

That means that while they may not be acidic to the taste or even particularly acidic on a pH test, they metabolize in our bodies into an acidic by-product. So naturally, most alkaline diet meal plans are going to be either vegan or vegetarian, or just very light on the meat and dairy. That doesn't mean, of course, that you need to completely remove meat from your diet, but if you choose to continue eating meat, you will need to be highly aware of the quality and quantity of the meat you consume. Keep it moderate and make sure to maintain good health otherwise and you should be okay.

But speaking of acid-forming diets, what so bad about them? We have seen some of the symptoms of having a seriously acidic blood pH level, but what if we're just prone to eating a little bit on the acid-forming side? Well, simply by virtue of the fact that we live in a modern world with modern luxuries and modern conveniences, most of our diets have strayed away from a good, healthy acid-alkaline blood balance, and maybe some of us without even knowing it is may be living with a chronic condition that results from such a diet known as 'chronic low-grade metabolic acidosis'.

This is what happens when our diet leans to the slightly acidic side for an extended, if not an indefinite amount of time. And the reality is that unless we take active steps to counteract this condition, it is highly likely that we will all succumb to it eventually, if not already. That is just the nature of the world we live in and the habits and practices of the industries and populations of our societies.

So if you are suffering from a form of low-grade chronic metabolic acidosis, perhaps without even knowing it, what are the signs? For one thing, you may notice some weight gain. This is a result of inefficient metabolic function on account of long-term low-grade acidosis. You may also suffer from pronounced but unspecific aches and pains. These will often be in the joints or even the bones. This type of pain associated with low-grade acidosis is likely the result of an acid buildup in the joints and bones.

Acid reflux, predictably, is also a good sign of this prevalent condition. But that is not the only part of your digestive system that can be affected. Long-term, low-grade acidosis can also cause a number of other digestive issues like intestinal cramping, irritable bowel, and generally poor digestion.

A whole host of other issues could manifest if you are one of the high percentages of people who unknowingly live day to day with a chronic case of low-level metabolic acidosis. Chronic fatigue and a general feeling of tiredness and muscle weakness may result. As can a number of other issues like skin problems, bone loss, kidney stones, receding gums, and urinary tract problems. So if you find that you have three or more of the many possible symptoms, then at least eighty percent of your caloric intake should be from alkaline-forming foods. The remaining twenty percent can more or less be of your choosing, but you may find high protein food items to be helpful.

A nice, quick and easy way to boost your body's alkalinity is by drinking beverages that are alkalizing. Spring water is one such naturally occurring source of alkalizing water. Also, water with a dash of lemon juice just as mentioned earlier. Green tea or ginger root tea will also have a similarly alkalizing effect.

You'll want to make sure that you are focusing primarily on eating whole foods. So that means vegetables and fruits, as well as root crops like potatoes and turnips. Also nuts and seeds can be an excellent alkalizing source and also a very strong source of protein. Beans are generally a good choice as well, although lentils, in particular, are renowned for their excellent alkalize boost. And when consuming grains, just remember that it is always best to consume whole grains.

Whether you are pursuing an alkaline diet to target a specific issue, or if you just want to have the healthiest body you possibly can, you are going to want to eliminate as much processed and artificial foods as possible. In fact, that goes for everyone, no matter what. Processed and artificial foods are doing anybody any favors, but they will certainly cause your body's blood pH level to lean to the acidic side. Refined sugars and added sugars rank very high up on the list of these types of foods that should be avoided at all cost. And refined white flour isn't doing you any favors either. In fact, some say it can be just as bad as refined sugars.

And while we are on the topic of eliminating things, if you can handle giving up coffee or any other caffeinated beverages, you will be giving yourself a major advantage on the path to a balanced blood pH level.

There are certain foods and nutritional elements that are acid-forming but that our bodies still need to function properly. These are things that we will have to keep a particularly close eye on in order to monitor amounts of intake. This includes essential fats, as well as pasta and other grains. If you are choosing to continue eating meat, then it should also be noted that meat and fish should both be consumed very sparingly and should also be very closely monitored and limited.

Finally, when it comes to dressing up your greens, namely when being consumed as a salad or being cooked, make sure to use high-grade and healthy fats like extra-virgin, cold-pressed olive oil, avocado oil, and coconut oil. All of which bring along tons of health boosts and benefits in addition to their alkaline-forming properties.

Chapter 7: Alkaline Meal Ideas

Now that we are fully informed and equipped to make good nutrition choices in regards to acid-alkaline blood balance, it is now time to put together some meal plans so we can put all of what we have learned into practice. We are going to look at a prime example of everything you might eat in a day to get the most out of your body.

This particular example isn't about limiting calories of eliminating any particular foods or food groups, so if you have any particular calorie counts you need to stay within, or if you are eliminating certain food groups from your diet such as meat or dairy, you may have to adjust accordingly. Just make sure that if you're replacing anything, to replace it with something of a similar acid-alkaline value, and that it serves that same food role as the replaced item. That is to say, replace proteins with proteins, carbs with carbs, et cetera.

And if you are not calorie counting, obviously use your best judgment here but there is no limit to how many alkalizing foods in the fruits and vegetable categories you can eat. Certainly, you should be limiting acidic foods like meats, dairies, grains, and processed foods if for no reason than to keep your acidity levels down, but fruits and vegetables especially the ones that are particularly alkalizing, you can eat to your heart's content.

So with that said, let's take a look at what our alkaline diet morning might look like.

We wake up nice and early in the morning, refreshed and ready to tackle our day because our healthy, alkaline-rich diet is allowing us to get the sleep we need and preventing our bodies from feeling overly fatigued. The first thing we do is get an ice-cold glass of water, squeeze some fresh lemon juice into it and drink the whole thing as fast as we can. As fast as we can without getting a brain freeze that is!

So now we're going to want to have a nice satisfying breakfast. Today, we are going to do a veggie scramble. Sounds great, doesn't it? This breakfast is going to consist of one or two eggs that we are going to scramble up with green onions, spinach or bok choy or any other leafy greens, and then some chopped bell peppers and diced tomatoes. You can even try it as an omelet if you like. Or even an egg-white omelet if you're feeling really healthy.

Or better yet, if you really want to go healthy, why not try that same breakfast, but as a tofu scramble instead of scrambled eggs. It's easy and delicious. Just replace the eggs with a handful of diced, firm tofu. You can season your tofu however you like, but we recommend trying a chili-style seasoning for some nice, tex-mex style breakfast burritos. Wrap optional.

After that amazing breakfast, we're going to have a nice productive and active morning. If we feel the need for a snack before lunch, we'll have maybe a fruit like an apple or a pear, maybe a banana or a handful of nuts or seeds. An ideal choice would be pumpkin seeds or almonds.

If you're anything like us, that already sounds like an amazing and healthy, nutritious day, but we haven't even gotten to lunch yet. So what might we enjoy for our midday meal on this ideal alkaline diet day? Why limit ourselves, let's look at a couple of options.

For one, we could try some lentil soup. This packs a nice alkalizing punch as it is but combine that with some steamed green like broccoli, carrots, onions, or kale, and you've got a powerful meal. Heck, why not steam up a mix of all of those veggies. Try a light olive oil-based salad dressing on the steamed veggies for some extra flavor.

As delicious as that sounds though, we still have another option. If you're still of the animal eating persuasion, you could try a nice big salmon steak, still a far healthier choice that its terrestrial cousin, served with some mixed greens which could include cucumber, carrots, tomatoes, and broccoli, among pretty much any other fresh veggies you would like. Similarly, you can season that with a light vinaigrette of your choice, but we particularly recommend a lemon and dill based one.

After all that amazing nutrition, you must be ready for a snack! Fight off that mid-afternoon slump with a nice alkalizing snack. This one you can keep nice and simple. Try a simple hard-boiled egg, seasoned with sea salt and fresh ground pepper to taste and a garnish of your choice if you're feeling fancy. Or if you're not inclined toward the animal-based foods, try a straightforward but delicious snack consisting of strips of sweet bell peppers, celery, or carrots, or a mix of all is always an option!

Finally, it's time for dinner and this is where you remaining meat eaters are going to have your way. You can have up to four ounces of your favorite meat, whatever that might be, but we highly recommend that if you must have meat, try to stick to something along the lines of fish, chicken, or other types of light poultry. You can serve this with a side of yam or sweet potatoes, baked or prepared in your favorite way and a nice simple garden salad with mixed greens and a light dressing of your choice.

Or for the plant-based folks, you can indulge in some pasta, but try to find or make pasta made from rice or quinoa, or other grains than wheat. Then you can top your pasta with all kinds of delicious veggies like broccoli, zucchini, and garlic. And then garnish with some olive oil and salt and pepper. Now you've had yourself a fresh and healthy food day!

PART 3

Acid reflux is one of the most common ailments affecting adults today. This is due to the notion that there are so many contributing factors and catalysts that can spawn its symptoms. As a result, educating oneself on the best ways to care for the symptoms of reflux is paramount for avoiding and overcoming them once they arise.

Evidently, a common refrain for most reflux sufferers is to avoid foods and drinks that are spicy and or fizzy. While these are certainly helpful approaches to take, you are well advised to consider approaches such as increased exercise for maintenance of overall health, avoiding certain exercises altogether if you are already suffering from severe GERD, implementing dietary changes such as limiting coffee intake, cutting back on peppermint and avoiding highly acidic foods.

But acid reflux symptoms are not caused solely by poor dietary choices alone. Indeed, smoking in excess and excess alcohol consumption are also consistent contributing factors that can induce acid reflux symptoms leading to GERD and Barrett's disease. Leveraging the knowledge presented in this book with regard to the specific dietary and lifestyle changes you must implement, along with acquiring a deeper understanding of the particular scientific reasons for acid reflux, will allow for an informed perspective so that your approach to the symptoms presented is equipped with the most up-to-date and effectual information.

When equipped with these strategies and general knowledge concerning acid reflux, you will certainly be at a distinct advantage when you are confronted by the painstaking and, at its most extreme, life-threatening, symptoms of acid reflux.

Chapter 1: Chronic Acid Reflux & Its Serious health Implications

Acid reflux is the result of abundant backflow of acid for your stomach into the esophagus. Anatomically, when your lower esophageal sphincter (LES) becomes weakened by, among many causal factors, continually consuming a high diet, acid can flow back into your esophagus causing acid reflux. Consequently, there are numerous symptoms that are spawned by acid reflux including, most notoriously, heartburn and indigestion. While high acidity in the gut is common for everybody and is often devoid of serious health concerns if occurring on a minimal basis, serious health concerns can develop if high acidity can persist for a prolonged period of time. If left unattended and without measures of control put in place, in many instances, these health issues will lead to hospitalization or even death.

In the most serious cases, chronic long-term heartburn, known as Gastroesophageal Reflux Disease (GERD), Barett's esophagus and esophageal problems can arise due to uncontrolled and unaddressed chronic acidity. Some of these most serious ailments are the remnants of years of neglecting the symptoms of acid reflux, particularly after big meals. Over time, the backflow of acid from your stomach damages your esophagus causing erosion of the layers lining the walls of the organ. This inflammation often leads to very painful swelling in esophagus called esophagitis and is accompanied by a painful swallowing feeling. In addition, esophageal ulcers are the most common ailment of an inflamed esophagus. In fact, GERD is the main cause of these ulcers in the lining of the esophagus. Moreover, painful swallowing, nausea, chest pain contribute to a lack of sleep which perpetuates many of these symptoms simplify virtue of that fact that your body is not getting the required rest to overcome these symptoms. As soon as these symptoms arise, be sure to consult a doctor too so that you can be prescribed medication to treat the symptoms before they grow into more serious, persistent conditions. Acquiring medial intervention is especially important when dealing with an ulcer; indeed, ulcers are extremely harmful to the lining of your organ and can be incredibly persistent if left unattended and given time to grow.

Even more, if your highly acidic diet remains the same as before the ulcer, this contributing factor will only enhance the growth of your ulcer and some of your other symptoms. Over time, the scarring of the lining within your esophagus will lead to scar tissue build up, thereby narrowing the esophagus altogether. Swallowing food and drinks will be made much more difficult as a result and may require a surgical procedure to expand the esophagus. These narrow areas of your esophagus are called strictures, and will likely lead to dangerous weight loss and or dehydration. Avoiding a procedure through immediate medical treatment of high acidity is clearly the preferred approach; but once these strictures block your esophageal pathways, a surgical procedure will be required.

Another common ailment of high acidity that affects many people is Barret's esophagus. Specifically, around 10%-15 of people who suffer from GERD will begin to also develop this painful condition, which results in dangerous changes in cells due to excess stomach acid. Thankfully, less than 1% of those who suffer from Barret's esophagus will actually develop esophageal cancer. If intervened early enough in the process, doctors are able to remove any of these abnormally developed cells through a procedure known as an endoscopy, whereby doctors will insert a flexible tube accompanied by a small camera into your esophagus. However, those who have GERD are at a, albeit slightly, increased risk of developing esophageal cancer. Even still, be sure to consult a doctor as soon as your symptoms reach a persistently painful level so that proactive measures and treatment can be implemented to quell your pain, and inhibit the growth of cancer cells.

If there is a long history of esophageal cancer in your family, you will be at an increased risk of developing this cancer as well, especially if you attain medical treatment after a prolonged period of time experiencing symptoms of high acidity. If you are aware of having a family history of esophageal cancer, make sure to ask your doctor for a regular endoscopy to find and mitigate the growth of improper cells. Moreover, tooth decay is a very common symptom of excess stomach acid as it wears down the outer layer of your teeth (enamel). As a result, this can lead to excess cavities and weakened teeth. In a recent study, researchers found that over 40% of GERD patients showed significant tooth decay, along with 70% of patients whose reflux had managed to reach the upper esophagus), compared to only 10% of those patients that had no symptoms of reflux. Certainly, reading about the symptoms that accompany stomach acid is disconcerting and worrisome. Nonetheless, being aware of the symptoms and knowing the early signs of their emergence will mitigate the risks of cancer and other chronic ailments like GERD and Barret's esophagus.

So, what can you do about these symptoms and potentially life-altering health concerns if you are experiencing excess stomach acid? Initially, and rather obviously, begin by assessing your diet. If you typically consume large meals, cut down your portions by at least 25%, and avoid eating right before you sleep. The latter is especially important for your digestive track as your body has to work harder to digest food whilst you are asleep and when your body's organs are supposed to be at rest. Also, limit your chocolate and coffee intake. Typically, medical professionals recommend limiting your coffee intake to only 2-3 cups per day at the most. If these levels are exceeded, your body's acidity levels will climb to healthy proportions and heighten your likelihood of acquiring the aforementioned symptoms. In the same way, excess consumption of alcohol and peppermint carry very harmful side-effects. Not mention, smoking is by far the most dangerous to many of your body's organs, especially the esophagus. Taking steps to cut back and eventually quit smoking is strongly recommended to not only avoid esophageal cancer and weakened tooth enamel, but for a bevy of other health-related reasons not directly associated with stomach acidity. When consulting a doctor to address stomach acid symptoms, you will most likely be prescribed an antacid, H2 blocker or a proton pump inhibitor (PPI); all three are available by prescription as well as over the counter.

Where GERD is concerned, there are other major factors that you should look out for. These symptoms are not specific to a particular body type or even certain foods. First, heartburn is, as mentioned, a clear indicator of GERD and is usually an only an occasional issue that is known to affect over 60 million people at least once or twice a month on average. However, for the 20 million individuals who suffer from heartburn on a chronic level through GERD, seemingly unrelated symptoms can inevitably result in numerous other health complications. You are well advised to consult a doctor if you find that you suffer frequent heartburn (two-three times per week regularly).

When you suffer from GERD, acid, food, as well as digestive juices tend to flow back into your esophagus from the pit of your stomach. Over time, this results in esophagitis, thereby leaving the king of the walls of the esophagus extremely vulnerable to additional harm through scarring, tearing and even deterioration. Additionally, while the primary symptom of GERD is heartburn, there are likely to be other symptoms that are far more difficult to diagnose for doctors and patients alike. Notably, doctors refer to a symptom known as, "silent reflux," which includes voice changes, chronic coughing, major and prolonged throat soreness, along with hoarseness. Patients may have a sustained sensation of having a lump in their throatier having the constant urge for having to clear one's throat. Another common symptom of GERD is the effect that stomach acid has on your breathing. Indeed, GERD, for instance, can heighten the extreme effects of asthma and or pneumonia. Whether or not patients have a history of lung problems personally or with regard to their family lineage, GERD can cause difficulty in breathing and persistent shortness of breath. However, treating this particular symptom is especially tricky; according to several recent studies, GERD medication, like PPIs, have been shown to increase pneumonia by directly contributing to the growth of harmful bacteria. Also, researchers have found that many prescribed PPIs suppress coughing that is needed to clear the lungs. As a result, your doctor may be forced to consider the function of your lungs when prescribing PPIs when in the process of treating symptoms associated with acid reflux.

Many people with ulcers from acid reflux tend to spit up blood and or see it in their stool. For a point of clarity, be sure to note that Esophageal ulcers are much different than stomach ulcers as they (stomach) are usually due to bacteria. Blood from esophageal ulcers, however, tend to be red or a darker purple-red color. If you find yourself having such symptoms, be sure to contact your doctor immediately. The immediate response from your doctor will likely be a schedules endoscopy mentioned earlier. In addition, you may also be prescribed acid-blocking or acid-reducing meds can treat these dangerous stomach ulcer.

An overarching common symptom of acid reflux is a lower quality of life. According to a 2004 study from Europe, whereby 6,000 GERD patients reported that their quality of life had been significantly diminished due to problems that are associated with drink, food and sleep, along with social and physical limitations. Not to mention, there can be major financial implications from having to buy an abundant amount of medications to treat the myriad symptoms of acid reflux, as well as the possible surgical procedures and endoscopy sessions that may be needed if the symptoms escalate to advanced stages. Moreover, the quality of life for patients of GERD was strikingly similar to heart-attack patients and was even lower, in certain cases, for those patients struggling with diabetes and cancer.

Generally, the healing time-frame for GERD is around 2 to 8 weeks. If allowed to persist without medicinal intervention, symptoms of GERD can inflict a considerable amount of damage. For example, reflux esophagitis (RO) can create visible and painful cracks and breaks in the esophageal mucosa. In order to fully and effectively heal RO, acid suppression for a prolonged period (roughly 3-9 weeks) is required and will likely be the timeframe advised by your doctor. Keep in mind that healing rates will rapidly improve as acid suppression increases.

Chronic stress is also a significant factor in the development, growth and. persistence of acid reflux. Our digestive system, moreover, is intricately associated with our nervous system. When stress presents itself, especially in an overwhelming or uncontrollable manner, our digestive system will then receive a lower amount of blood flow, thus causing various issues. Further, our gut bacteria are implicated in our management of stress at increasing levels, so probiotics are helpful in helping the management of this development.

Chapter 2: The role of Fibre, Prebiotics and Probiotics

Incorporating key changes into your diet can carry massive benefits with regard to dealing with the many symptoms of acid reflux. Specifically, consuming more fiber is an excellent way to mitigate the harsh symptoms. An important point of clarity is the distinction between dietary fiber, defined as edible but non-digestible carbohydrate-based material, and insoluble fiber. Dietary fiber is mainly available in abundance naturally in many cereals, grains, plants, and vegetables as these all play a major role in gastrointestinal health. Given the importance of fiber, and its positive impact in easing the symptoms of acid reflux, many people on average are dangerously deficient in fiber. This deficiency includes both the soluble and insoluble forms of fiber. The main difference between these two forms of fiber are found in their role in digestion; insoluble fiber expedites the travel of foods through the gastrointestinal tract, while solute fibers have been shown to slow the digestion process.

While the varied and, at times, monotonous science concerning acid reflux is still in progress and remains to be settled. Nevertheless, the theoretical benefits of adequate intake of fiber include avoiding trigger foods altogether, as well as the stomach-filling "full" effect of fiber and fewer relaxation reactions of the anti-reflux valve residing between the stomach and the esophagus. Even still, there exists a persistent relationship between acid reflux trigger and the role of fiber. To elaborate, soluble fiber has been shown to induce the body to draw fluid out of already digested food, which then contributes added bulk to your meals which leaves you with a feeling of being "full" for a far more prolonged period of time. As it is commonly found within such sources as barley, peas as well as oat bran, soluble fiber does play an active role in regulating glucose levels and may even contribute to signaling the brain that the stomach is in fact full both during and after eating a meal of any size. Moreover, smaller meals can help acid reflux by refraining from overfilling your stomach. Whereas, insoluble fiber as found within vegetables and whole grains can speed up the passage of stomach contents to your intestinal tract, thereby decreasing your body's propensity   for   reflux. Additionally, fatty, fried foods are typically much lower in fiber and are also frequently accompanied by the triggering symptoms of regurgitation, indigestion, and heartburn. Indeed, a fiber-rich diet like fresh fruits, vegetables, and whole-grain bread can tend to contribute to fewer instances of reflux symptoms arising. Some dietary fibers are also widely considered to be probiotics. Note the key distinction between prebiotics and probiotics: the latter refers to the specific helpful bacteria itself, with prebiotic referring to bacterial nutrients. In other words, prebiotics are nutrients which are left for bacteria to digest, or, more plainly, fuel to encourage the balanced bacterial growth within digestive organs. Here, the role of fiber in greatly improving the many symptoms of acid reflux is illuminated; this role is as a bacterial intermediary. On the whole, nonetheless, particular items in your diet tend to perform a seemingly minor role in the symptoms associated with acid reflux. Those who suffer from chronic acid reflux are strongly advised to avoid those specific foods that can aggravate painful heartburn and regurgitation; however, eliminating a broad range of food from your diet is no recommenced as it can detrimental to your overall health. Instead, you should note foods and beverages that can trigger acid reflux specifically and root them out from your diet as soon as possible.

There is a tremendous amount of evidence in favor of incorporating a fiber-rich diet into your daily routine. Notably, the benefits are particular to the overall maintenance of your gut and with regard to managing the amount of harmful bacteria native to that region of your body. A study from 2004 that involved over 65,000 people revealed that fiber intake was associated with the improved perception of acid reflux symptoms. Also, this study revealed people who consumed high fiber bread were nearly three times as likely to experience relief of acid reflux symptoms compared to people who consumed bread with lower fiber content. Granted, the reasons for these results remains unknown; however, the authors of the broad study have speculated that the digestive process of fiber can also be a catalyst for enhanced muscle relaxation from the stomach through the esophagus as it tightens the anti-reflux valve.

So, what are the disadvantages of fiber in acid reflux? It is true that fiber is especially helpful when serving to ease acid reflux symptoms, excess fiber consumption has been shown to aggravate the symptoms. A study published in a medical journalism 2014 indicated that consuming a minimum of 10 grams of highly fermentable starches each day can significantly contribute to painful episodes of acid reflux symptoms. An additional study noted that nine participants who were diagnosed with GERD, those patients who consumed a prebiotic known as fructooligo also had elevated instances of acid reflux symptoms than those patients who were given a placebo.

There are many aspects to consider when striving to efficiently manage acid reflux symptoms. Among them, dietary fiber is perhaps the most important or, at the very least, the most consequential. For instance, while being excessively overweight is certainly risk factor as far as GERD is concerned, adequate consumption of healthy fiber will aid in keeping the weight at a healthy level. Excess fiber, however, causes stomach distension in may people, along with enhanced stomach pressure as well as prolonged emptying on the stomach in many cases, all of which have shown to lead to accentuated acid reflux symptoms. Medical professionals specializing in gastroenterology strongly recommend implementing lifestyle changes such as eating smaller sized meals on a frequent basis (as opposed to larger meals a few times per day), limiting overall consumption of carbonated beverages and foods high in salt content with the intent of improving the acid reflux symptoms. Note that women over the age of 50 should try to consume 25 grams per day; on the other hand, men under 50 are strongly advised to consume 38 grams of fiber each day. While consuming more laxatives is often the approach undertaken by people dealing with symptoms of acid reflux, medical professionals advise increasing your fiber content within your diet for maximized results, as well as maintaining your overall health as laxatives can take a substantial toll on your body and its digestive tract. Consuming a higher amount of fiber will strengthen your stool, keep wastes traveling more smoothly through your intestinal tract, along with preventing constipation. During this process of consuming more fibre, be sure to also ensure that you are consuming plenty of water as well; for fiber to have its absolute best effect, it is imperative that your body remain as hydrated as possible to make sure that waste moves smoothly along your intestinal tract rather than building up due to rigidity.

A good source of probiotics and fiber is yogurt, which carries "good" bacteria helpful for overall maintenance of gut health. A healthy gut is paramount for an efficient and well-functioning intestinal tract and digestive system. Prebiotics and probiotics are, essentially, analogous to food for the bacteria in your stomach; bananas, corn, and whole wheat are additional food sources that are high in prebiotics. Moreover, one of the most beneficial aspects of a fiber-rich diet is the notion that high-fiber foods help control your cravings for snacks. Certainly, high fiber diets can help you lose weight as it displaces other calories for overall maintenance of health.

Guarding against illness, fiber-rich diets will also lower your chances of developing major gut-related illnesses such as diverticulitis. This condition, pockets in the walls of colon trap waste as opposed to moving it along. While doctors remain unsure of the direct catalyst(s) for the illness, consuming a high-fiber diet moves waste fervently along through your system. Along with diverticulitis, a high-fiber diet also eases and prevents irritable bowel syndrome— which has also been linked to acid reflux, albeit a rarer, more extreme symptom. Nevertheless, the most common symptom of acid reflux— heartburn— is quelled by a fiber-rich diet.

Probiotics are becoming increasingly linked to the management of the symptoms associated with acid reflux and alleviating these issues. For a more technical explanation, as we are already aware that probiotics are an effective way of balancing the gut bacteria inside of our bodies, they also help combat against a bacterial infection knows as, H. pylori. This bacterial infection usually originates in childhood. This bacteria, found in the stomach, can alter the environment around them through reducing the acidity levels so that they can survive for longer periods of time. By penetrating the lining of the stomach, thereby remaining hidden and protected by the mucous membrane so that the body's immune defenses cannot reach them. In addition, these bacteria tend to secrete an enzyme called urease, which converts urea to ammonia. The presence of ammonia in this instance is significant because it reduces the stomach acidity around the specific area where the bacteria is found. Coincidentally, it is this lower stomach acid that is often mistaken— by doctors and patients alike— for acid reflux. Can probiotics help combat H. pylori? Well, many medical researchers believe that probiotics can, in fact, help battle this bacteria in several key ways. For one, probiotics are believed to strengthen the protective barrier against H. pylori by producing antimicrobial substances, along with competing against H. pylori for what is known as adhesion receptors (space on the lining of the stomach). Also, it is believed that probiotics assist in stabilizing the gut's mucosal barrier. Many researchers even argue that the production of relatively large amounts of lactate is another inhibitory factor of H. pylori due to the possibility that it could lower the H. pylori urease. Not to mention, probiotics may also be effective in modifying inflammation levels by interacting with the epithelial cells that are responsible for managing the secretion of inflammatory proteins in the gut.

Depending on the particular cause of your acid reflux, probiotics can be incredibly useful for alleviating the painful symptoms. Probiotics can be taken in conjunction with an antacid without worry of the antacid overtaking the positive benefits of the probiotic. More importantly, your approach should be to uncover the root of your acid reflux and adjust medical intervention accordingly. Of course, you doctor plays a huge role in this process, especially in diagnosing the specific cause of your acid reflux; still, it is your responsibility to seek medical expertise so that you can tackle the root of the symptoms as opposed to addressing the individual symptoms as they arise. Being aware of the triggers in both the food and drinks that you consume, managing your stress levels regardless of how often they fluctuate, and finding the specific levels of acid within your stomach are critical steps to address the symptoms of acid reflux. More likely than not, the best method of dealing with your symptoms will be to implement a diverse approach that incorporates a range of approaches rather than relying solely on medical intervention.

Chapter 3: Understanding the role of proteins, carbs, AND fats in healing acid damage

In recent years, there have been increasing reports concerning the benefits of a low-carb diet in healing the damage induced by acid damage. This may seem counterintuitive given the notion that the standard treatment for GERD includes the removal of certain foods that increase acidity in the stomach, for example, tomato sauces which are believed to be contributing causes of excess stomach acidity. Also, as mentioned in chapter 1, the removal of coffee, alcohol, smoking, and peppermint are other dietary and lifestyle changes that ease GERD symptoms. Additionally, researchers have found that diets with a higher amount of carbohydrates can significantly elevate symptoms of acid reflux. Whereas, a low-carb diet has also been shown to reduce symptoms of GERD. While many health researchers and medical experts have expressed concern over the exceptionally low proportion of daily calories from fat and protein in low-carb diets, as wells calorie levels being considerably lower in these diet than recommended. Nevertheless, the effects of gastroesophageal reflux disease have been shown to be significantly reduced after implementation of a low-carb diet. For a case in point, following a 2001 research study in which 5 individuals with diverse ranges of GERD symptoms and across the age spectrum, engaged in a low-carb diet, each of the 5 research participants showed significant relief of symptoms. Granted, throughout the duration of the study, which spanned 8 months, 3 of the 5 research participants also reduced their coffee intake concurrently.

The concurrent reduction of coffee, coupled with lower intake of carbohydrates, was shown to be effective in reducing symptoms such as heartburn and stomach pain. Interestingly, while coffee reduction was a contributing factor, observations from a few of the participants revealed that exacerbating foods such as coffee and fat are less pertinent when a low-carb diet is strictly followed. In other words, when implementing a low-carb diet, the effects of classic factors like coffee and fat intake are vastly diminished even if their consumption is not significantly reduced. So, whether or not you choose to reduce coffee and fat intake significantly, you are likely to reduce most GERD symptoms by solely undertaking a low-carb diet. Therefore, the logical conclusion to draw from these findings is that a low-carb diet is a significant factor for reducing the symptoms of acid reflux. This conclusion gains even more credibility when considering the propensity for high-carb diets to aggravate GERD symptoms.

In addition to lower carbohydrates, lean sources of protein and healthy fats are beneficial for reducing symptoms associated with acid reflux. Lean proteins found in eggs are a great addition to your diet for reducing acid reflux symptoms; however, they are a problem for many people due to elevated cholesterol. If eggs are an issue for you, be sure to consume only egg whites and refrain for consuming higher fat yolks—which have been shown to elevate GERD symptoms. Moreover, as high-fat meals and fried food typically descries LES pressure thereby delaying emptying of the stomach and boosting the risk of acid reflux, it is in your best interest to choose lean grilled meats, as well as poached, baked or broiled meats. Boosting proteins, in the way, will also provide benefits for your overall health as well. Also, complex carbohydrates, as found in rice, whole grain bread, couscous and oatmeal carry excellent benefits for reducing GERD symptoms and easing any scarring that may already be present in the walls of your esophagus. Specifically, brown rice and whole grains add valuable fiber your diet. Root vegetables such as potatoes are excellent sources of healthy carbohydrates and easily digestible fiber. Remember to refrain from incorporating too much garlic and onion while preparing your meals, which are can commonly irritate the esophagus and stomach lining.

Along with proteins and complex carbs, incorporating healthy fats has great benefits for easing GERD symptoms and other symptoms accompanying acid reflux. A type of nutrient, fat is high in calories but certainly a necessary component of your diet. Keep in mind, however, fats can vary and they do not all have the same effect on your body. On the whole, you are well advised to avoid consuming a high amount of saturated fats as typically found in dairy and meat, along with trans fats found in processed foods, shortening and margarine. As a replacement, unsaturated fats from fish and or plants are recommended; some examples of monounsaturated fats include sesame, olive, canola, sunflower, peanuts, seeds and nuts, as well as butter. In addition, examples of polyunsaturated fats include such oils as safflower, corn, flaxseed and walnut, fatty fish such as salmon and trout, along with soybean.

Some other helpful tips for reducing acid reflux symptoms include chewing gum, as long as it is not spearmint- or peppermint-flavored. Chewing gum increases the amount of saliva production in your mouth and also reduces the amount of acid in your esophagus. While alcohol consumption has already been mentioned in earlier chapters, research suggests that some people begin to experience extreme symptoms after only one drink; if you fall into this category, be sure to experiment with your levels to uncover what amount is best for you. Additionally, during and after each meal, particularly bigger meals, be sure to remain aware of your posture. Generally, it is a good idea to sit up while you are eating and avoid lying flat on your back for at least hours post-meal. Standing up and walking around the room after a big meal can help encourage the flow of gastric juices in the right direction. Further, avoiding eating big meals before bed can help you refrain from overloading your digestive system while you sleep. Digestion increases the overall amount of gastric acid that is present within your stomach. When laying down, LES's ability to inhibit stomach contents from traveling through the esophagus decreases significantly. When operating concurrently, excess stomach acid and remaining in a reclined position for an unexpended period of time create an abundance of acid reflux symptoms. On the whole, consuming a large meal for less than 3 to 4 hours prior to bed is generally not advisable for those suffering from persistent GERD symptoms; however, the timing of these symptoms can certainly vary depending on the individual.

In a 2017 research study on the benefits of healthy dietary changes versus drug intervention, researchers studied the effects of dietary changes to a type of reflux known as laryngopharyngeal reflux or LPR. This reflux is essentially triggered when pepsin, a digestive enzyme from the stomach, reaches the sensitive tissues in the upper section of the digestive tract.

Symptoms like throat clearing and hoarseness are common with pepsin in the throat and or upper part of the digestive tract. In the study, the researchers had participants suffering from acid reflux switch to a Mediterranean diet and consuming significantly more water, thereby neutralizing excess acid. In this particular study, participants avoided classic triggers such as coffee, peppermint, alcohol, fatty and spicy foods, chocolate and soda. Another set of participants were given pharmaceutical drugs to ease GERD symptoms.

After a six week timeframe, participants of the study reported a greater percent of declines in their GERD symptoms as those participants that had used drugs to address the symptoms. Granted, the study did not elaborate on the particular ways in which that the diet and increased water consumption eased the symptoms; nonetheless, the Mediterranean diet incorporates eating mostly plant-based fruits and vegetables. In addition, the increased water can mitigate pepsin's acidity levels in the throat. As mentioned, fruits, vegetables and water are great methods of reducing acid reflux and GERD symptoms. With this in mind, the positive benefits experienced by the study's participants is not surprising in the least.

Adjusting your diet for GERD does not require removing all of the foods that you typically enjoy eating. Instead, a few simple changes to your diet if more than enough to address the uncomfortable symptoms of GERD. Your aim in addressing the GERD symptoms should be to create a well-rounded, nutrient-based diet that incorporates a variety of foods that include vegetables, fruits, complex carbs, healthy fats and lean sources of protein. Healing acid reflux damage is made significantly easier when starting with dietary changes that add healthy and diverse foods. Coupled with medical intervention (if required), healthy dietary changes can carry great benefits for healing scarring in your esophagus and stomach, as well as symptoms such as heartburn, bloating and even tooth decay.

Chapter 4: Exercise to reduce acid reflux

When GERD symptoms escalate and you are still in the early stages of implementing dietary changes to address acid reflux symptoms, exercise can be a great option for reducing the symptoms. When GERD symptoms begin to arise, high-impact physical activities like running, skipping rope and jumping exercises. If you are overweight or obese, a weight loss of 10% has been shown to reduce GERD symptoms like heartburn, bloating and reflux. A self-reported analysis study of individuals experiencing GERD symptoms fund that those who reduced their Body Mass Index (BMI) by 2 kilograms or 4.4 pounds or more, along with the circumference of their waist by 5cm or more has improved their GERD symptoms significantly. In contrast, there are also certain exercises that can induce reflux by opening the lower esophageal sphincter (LES) during workouts such as heavy lifting, stomach crunches, or other high impact exercises. When the LES opens, stomach acid travels up into the esophagus causing heartburn.

There are some common sense tips concerning exercise for managing GERD symptoms. First, this twice about how much and what you are eating prior to starting your exercise routine. Obviously, less food in your stomach is ideal. If you are too full, you should wait at least 1 to 2 hours before initiating your exercise routine. This will allow for food to pass fro your stomach through to the small intestine. With less food in your stomach while exercising, it is significantly less likely that you will experience the painful and annoying symptoms of acid reflux such as heartburn and bloating. Next, you should choose the food you consume with thought and, in some instances, caution. Generally, you should avoid foods that trigger GERD symptoms (choosing complex carbohydrates is advised). Your stomach does metabolize these foods much faster than others through a process known as gastric emptying. Moreover, diabetics should avoid high fat and high protein foods before exercise due to being more susceptible to experiencing slow gastric emptying. Experts also suggest adjusting your workout if you suffer from frequent GERD symptoms. Starting at a slower pace with workouts that put less strain on your body like controlled walking and controlled weight training in either a sitting or standing position is strongly recommended. Whereas, high impact, high-intensity workouts like running, rowing and cycling and induce acid reflux. Additionally, acrobatic workouts and gymnastics can also disrupt stomach contents. The key is to avoid exercises that jostle the LES and reflux, these are typically positions that put your body in awkward positions like being upside down defying gravity in one way or another.

A great exercise that carries tremendous benefits for improving and relieving acid reflux symptoms and digestion is yoga. One particular study from 2014 found that six months of yoga significantly reduced acid reflux and stomach bloating, along with improved esophagitis. Again, however, try to avoid positions that heighten GERD symptoms. If any of these "lifestyle" changes fail to improve your GERD symptoms during exercise, be sure to consult your doctor about being prescribed medication for acid-suppression. And, of course, engaging in a constant exercise routine is not only very beneficial with regard to improving your GERD symptoms, but also for the maintenance of your overall health.

Chapter 5: How to live a reflux free life?

As you may have already noticed, acid reflex can be induced by an abundance of factors ranging from diet, bad habits, poor sleep hygiene, and many other factors. Clearly, it isn't just as simple as cutting out bad habits and instilling a series of great dietary and lifestyle changes. But whether you are able to successfully implement these changes or not does not, thankfully, hinge on strict adherence to a stringent diet or eliminating some of your favorite guilty habits. But before you can begin to consider stepping into a reflux-free life, you should be cognizant of the stages of reflux and recovery.

Firstly, almost everyone who suffers from GERD begins with normal LES and little to no reflux. The severity level of GERD, therefore, more than likely correlates to best with the overall degree of damage inflicted upon the sphincter. Note, however, that this is not easy to determine. Normally, the amount of damage to your sphincter correlates with the overall severity of acid reflux symptoms. This severity is most often determined by the volume, frequency, and duration of reflux episodes. In turn, these factors will correlate with GERD symptoms such as regurgitation and heartburn. If you are diagnosed with GERD, your strategy for addressing the symptoms and eventually overcoming them should first be to containment. Unfortunately, damage to your LES caused by GERD cannot be reversed by drugs and is permanent. Nonetheless, many patients of GERD have been able to live with these symptoms and with functionality despite damage to the sphincter. Changing simple lifestyle habits, such as sleeping and eating, can significantly decrease and prevent severe reflux episodes in spite of damage to your sphincter.

In Stage 1 of GERD, known as Mild GERD, most adults currently have minimal damage to their LES and tend to experience mild GERD occasionally. Most often, the adults are left with either tolerating occasional heartburn or will have to use over-the-counter acid suppressive medications from the onset of symptoms through its subsequent stages. Typically, when taking drugs to address the symptoms, quality of life for patients is not affected because the medications are very effective in suppressing symptoms. If you choose to take medication to address symptoms, make sure that you are also cleaning up your diet. If you continue to consume trigger foods and beverages, like coffee and certain sauces, for example, the benefits you garner from medication will be minimal and your recovery will be prolonged if not inhibited altogether. Replacing these meals with smaller, leaner meals that do not pose a threat to inducing symptoms of GERD is recommended. This will ease heartburn and lessen the damage to your sphincter.

In stage 2, known as Moderate GERD, symptoms are far more difficult to control and use of prescribed acid-suppressive drugs will be needed. In this stage, reflux is accompanied by symptoms that are far more intense than stage 1. Therefore, medicinal intervention is needed to mitigate the damage and pain caused by these symptoms. Still, many symptoms in this stage can be managed by using acid-suppressive drugs for prolonged periods of time. Keep in mind that over-the-counter medication can provide insufficient relief; whereas, prescription strength medications are needed in order for you to manage GERD symptoms effectively. Additionally, stage 3, or, Severe GERD, can result in a very low quality if life and is generally considered to be an extremely serious problem by medical professionals specializing in GERD. Because prescription grade acid-suppressive drugs and medicinal intervention usually do not control symptoms, regurgitation is frequent. In Stage 3, it is entirely possible that complications associated with erosive GERD are present. Lastly, stage 4, or, reflux-induced esophageal cancer, is quite obviously the most serious stage. The result of numerous years of severe reflux, nearly 16% of all long-term reflux sufferers progress to this extremely advanced stage. Due to the long-term reflux, the esophagus' lining has been heavily damaged, thereby resulting in a high degree of cellular changes. Also, stage 4 is the stage that involves the pre-cancerous condition called Barrett's esophagus and or an even more severe condition called dysplasia. Granted, these conditions are not cancerous. However, they are capable of raising the risk of developing reflux-induced esophageal cancer. Accordingly, at this stage, common GERD symptoms are likely to also be accompanied by a strong burning sensation in the throat, chronic coughing, and persistent hoarseness. A narrowing of the esophagus, or strictures, will also be present in this stage, and can also be characterized by the sensation that food is sticking to your throat. However, this is only a feeling associated with strictures. Notably, this symptom is also caused by esophageal cancer. Keep in mind that stage 4 GERD can only be diagnosed by a medical professional through an endoscopy and from an intrusive biopsy of cells retrieved from the lower esophagus.

A 30-day recovery plan for GERD symptom can be easily broken into weekly steps. In the first week, you will presumably be trying to lean off of the medication that you have been prescribed. A reversion plan of this nature should take into account a variety of approaches that incorporate dietary changes, exercise routines, sleep schedules, and other lifestyle changes. Drinking more tea and water is strongly recommended throughout your 30-day reversion plan, as long as the tea is not peppermint-flavored. Also, in your first week of recovery, be sure to get as much sleep as possible, in conjunction with eating 2 to 3 hours before you sleep. A meal should also be incorporated throughout this entire 30-day stretch. This plan should include egg whites in the morning, instead of coffee switch to tea for your caffeine fix. For lunch, whole grain bread with lean meat- chicken or turkey preferably- with light sauce and a bevy of vegetables. Moreover, dinner should include a balanced meal that provides nutrients and foods that will not induce heartburn. Remember, this meal should be consumed a few hours before bed so that your body has time to properly digest the food. Also, be sure to refrain from lying directly on your back after your meal; as mentioned, this will induce acid reflux symptoms like heartburn and excess bloating.

In week 2, in keeping with the consumption of nutrient-rich food and water consumption from the previous week, you should begin an exercise routine of you have been devoid of routine prior to week 2. A consistent exercise routine will help maintain overall health so that your body has excess strength and energy to overcome the wear and damage inflicted from GERD symptoms. Also, a consistent exercise routine will boost your metabolism so that you can burn off excess unhealthy fats and complex carbohydrates that can cause strain to your body and induce reflux symptoms. In the third week of your 30-day revision plan, you can slightly increase the size of your meals. Still, make sure that the overall size of your meals remain relatively small, with only slight additions where you may see fit. By week three, your exercise routine, water consumption, and sleep habits should be starting to feel more routine and many of reflux your symptoms will begin to diminish. Moreover, leading into week 4, your diet should continue to incorporate nuts, vegetables, fruits, tea, and other plant-based foods and drinks to expedite the healing process. However, the final week of your 30-day recovery plan is vital for sustaining the progress that you have presumably made since the start of the month. It is vital because you must ensure that you do not get too comfortable in your routine that you allow for gradual decline back into the habits that spawned your acid reflux. Pushing through this final week will augment your progress and solidify your path to living a reflux-free life. Specifically, with regard to your diet, you can incorporate the following fruit and vegetable smoothie recipe. Smoothies and healthy shakes are an excellent meal replacement option for optimal health and recovery from acid reflux.

Add 2 scoops of frozen berries into a 400ml cup

Add 2 scoops of spinach from a 250ml scoop

Add 2 tablespoons of Chia Seeds

Add 2 tablespoons of hemp hearts

Add 1 tablespoon of peanut butter

This smoothie should be blended with water to ensure that it is not excessively thick.

CONCLUSION

Good job for making it through to the end of the book. We hope it was informative and able to provide you with all of the tools you need to achieve your goals whatever they may be.

The next step is to put everything that you have learned in this book into practice. Learning more about your body and how it works is always a great place to start and in this book, we learned all about the acid-alkaline balance in our bodies, how the chemistry works, and what are the effects this balance or imbalance can have on our health.

We learned about some of the cutting-edge science behind our understanding of pH levels in our body, and how we can fine tune them to the perfect level through our diet and other important practices. The human body is a very complicated and delicate machine, but the more we come to understand it, the better off we are and the more educated we can be in our choices regarding our health and wellbeing.

Given that your acid reflux is likely to exist on a ranging spectrum and will most definitely be different for every individual, this book provides methods of addressing your symptoms and ailments in a manner that is applicable and helpful for whichever stage of the reflux you are experiencing. What's more, the 30-day recovery plan is a great way to ensure that your symptoms are staved off for good once you can utilize medical intervention. If, for instance, you are prescribed drugs to mitigate your symptoms, be sure to use the methods outlined in this book in conjunction with the prescribed amount of medication that your doctor has instructed.

Where your health is concerned, being prepared and informed is critical to seeing that you successfully overcome the ailments that are affecting you.

That is precisely why this book is of value. After reading, you are now informed about the intricate aspects of acid reflux and should feel extremely prepared as you move forward in addressing these symptoms.

Now, it's up to you to take what you have learned in the preceding chapters and apply them into your life. Do you have what it takes to be in full control of your own health and get the very most out of your body? Will you settle for a body that works within tolerable levels, or do you want to maintain peak balance in your life and in your health? Then now is the time to apply all that you have learned and prove it for yourself!